Sharman Apt Russell teaches writing at Western New Mexico University and at Antioch University in Los Angeles, California.

Praise for
An Obsession with Butterflies

"This enchanting little book is an extraordinary mating of exciting, sure-footed science and inspired prose poetry. It abounds with perfect little declarations and revelations. If you have never given butterflies much thought, this will enrich your perceptions of the world around you. If you already are an addict, this book will mightily confirm your affliction."

—Baltimore Sun

"Russell has done a wonderful job of gathering up bits of lore to brighten up her account...all of which makes for fascinating reading."

—The Economist

"An exquisitely buoyant ode to butterflies...captivating blend of science, history and wonder...Fascinating."

—Boston Herald

"Russell merges wit, knowledge and poetic language in this engaging scientific rumination.... A well researched and beautifully written natural history of these colorful insects."
  —Publishers Weekly

"Russell speaks clearly and enticingly.... Highly enjoyable: a modest lepidopterous encyclopedia that piques and prods the reader into wanting to know much more."
  —Kirkus Reviews

"Readable, informative, and at times poetic."
  —Boston Globe

"One of the most elegant, beautifully written nature books I've ever read.... At last, butterflies have a work that is worthy of their beauty and grace."
  —Curled Up with a Good Book

"Unwraps the exquisiteness of the butterfly, revealing a winged creature worthy of the enchantment that inspired Apt Russell's book."
  —Portland Oregonian

"Lore and natural history come together here for a fine read."
  —Dallas Morning News

"Russell's writing is extraordinary."
  —Buffalo News

An
OBSESSION
with
BUTTERFLIES

*Our Long Love Affair with
a Singular Insect*

SHARMAN APT RUSSELL

A Member of the Perseus Books Group
New York

To the butterfly girls:
Faye, Lorrie, Maria, Rachel, and Brisly Alejandra

Many of the designations used by manufacturers and sellers to distinguish their products are claimed as trademarks. Where those designations appear in this book, and where Perseus Publishing was aware of a trademark claim, the designations have been printed in initial capital letters.

Copyright © 2003 by Sharman Apt Russell
Illustrations copyright © 2003 by Jennifer Clark

All rights reserved. No part of this publication may be reproduced, stored in a retrieval system, or transmitted, in any form or by any means, electronic, mechanical, photocopying, recording, or otherwise, without the prior written permission of the publisher.
Printed in the United States of America.

Library of Congress Control Number: 2003102851
ISBN 0-7382-0699-7 (hc); ISBN 0-465-07160-0 (pbk.)

Basic Books is a Member of the Perseus Books Group.
Find us on the World Wide Web at http://www.basicbooks.com.
Perseus Publishing books are available at special discounts for bulk purchases in the U.S. by corporations, institutions, and other organizations. For more information, please contact the Special Markets Department at the Perseus Books Group, 11 Cambridge Center, Cambridge, MA 02142, or call (800) 255–1514 or (617) 252–5298, or e-mail special.markets@perseusbooks.com.

Text design by Brent Wilcox
Set in 10.5-point Fairfield Light by the Perseus Books Group

First hardcover printing, May 2003
First paperback printing, April 2004

2 3 4 5 6 7 8 9 10—06 05 04

OTHER BOOKS BY
SHARMAN APT RUSSELL

Anatomy of a Rose
Songs of the Fluteplayer
Kill the Cowboy
When the Land Was Young

The beauty and brilliance of the insect are indescribable and none but a naturalist can understand the intense excitement I experienced. . . . On taking it out of its net and opening the glorious wings my heart began to beat violently, the blood rushed to my head, and I felt more like fainting than I have done when in apprehension of immediate death. I had a headache the rest of the day.

ALFRED RUSSEL WALLACE
The Malay Archipelago

CONTENTS

CONTENTS

ACKNOWLEDGMENTS

During the course of writing this book, my daughter, Maria Hallie Apt Russell, was a wonderful research assistant and companion to such places as England and Costa Rica. My husband, Peter Russell, was there from the beginning to the end, envisioning, re-envisioning, and catching butter-flies. My editor at Perseus Books, Amanda Cook, is always supportive, insightful, and on target.

Any mistakes in the following chapters are my own. At the same time, I must thank the many scientists whose research I used and who did me the favor of reading over the text.

Bert Orr from Griffith University in Queensland, Aus-tralia, was a tremendous help. A good writer and editor, as well as an expert in his field, Bert has the gift of offering critical suggestions in the context of generous enthusiasm. Some of his work with butterflies is highlighted in the chapter "Love Stories." His comments and observations strengthened many other parts of this book.

ACKNOWLEDGMENTS

Philip DeVries also offered suggestions and comments in the middle of a very busy life. At the Natural History Museum in London, Richard Vane-Wright, Phillip Ackery, Jeremy Holloway, and David Carter all took time to meet with me and later to read sections of the manuscript. I first used Martha Weiss's work in my book *Anatomy of a Rose* and was happy to return to her research. Dan Papaj at the University of Arizona offered valuable ideas and commentary, as did Larry Gilbert at the University of Texas at Austin.

Rudi Mattoni gallantly sat with me in the lounge of the Marriot Courtyard Hotel in Marina del Rey and told stories. I would also like to thank Arthur Bonner for his help.

Bill Toone of the San Diego Zoo had some important ideas about the business of butterflies. Daryl Loth shared some nice moments by the Rio Tortuguero. I am grateful for the hospitality of Paul Grant at the Barra del Colorado Wildlife Refuge field biology station and the good work of the Canadian Organization for Tropical Rainforest Education and Conservation (COTREC). Joris Brinckerhoff of CRES and The Butterfly Farm in San José, Costa Rica, also gave me his time and comments.

Two organizations I have to applaud are The North American Butterfly Association and The Xerces Society. More information can be found at their Web sites.

Finally, I would like to thank Western New Mexico University and the WNMU Research Committee for their ongoing support. In particular, I have to praise again the staff of the Interlibrary Loan Department, without whom nothing could be written.

A NOTE ON NAMES

Most people classify butterflies into the following groups:

The superfamily Papilionoidea are divided into five families.

The family Papilionidae are often called the swallowtails. They include the birdwings and the apollos.

The family Pieridae encompass what we call the whites, orange tips, brimstones, and sulphurs.

The family Lycaenidae include the blues, hairstreaks, and coppers.

The family Riodinidae are known as the metalmarks. Some taxonomists list this as a subfamily.

The family Nymphalidae are referred to as brush-footed butterflies because their forelegs are often reduced to a hairy pad. There are thirteen subfamilies that include these common names: owls, satyrs, longwings, morphos, browns, milkweeds, snouts, fritillaries, tortoiseshells, painted ladies, admirals, buckeyes, checkerspots, and crescents.

The superfamily Hesperioidea include only one family, Hesperiidae. These butterflies are commonly called skippers. They tend to have a quick, erratic flight and a short, stout body. Skippers can resemble moths in the way they hold their wings out and their use of cocoons for pupation.

The superfamily Hedyloidea also include only one family, Hedylidae. These "butterfly-moths" have characteristics of both moths and butterflies.

For the most part, I use common names for species. I do not capitalize general terms such as fritillaries or whites; I do capitalize the common name referring to a specific species, such as a Silver-washed Fritillary or a Large White. Of course, common names bedevil scientists and are ridiculously inaccurate because they change with locale and custom. For that reason, I have included an alphabetical list of every common name used in the text, along with its scientific name. For non-lepidoptera, the scientific name is given in the end notes.

anise Swallowtail

LEPIDOPTERA

Acmon Blue *(Pliebejus acmon acmon)*
American Painted lady *(Vanessa virginiensis)*
Apollo *(Parnassius apollo)*
Australian Big Greasy *(Cressida cressida)*
Baltimore Checkerspot *(Euphydryas phaeton)*

Black Satyr *(Satyrus actaea)*

Black Swallowtail *(Papilio polyxenes)* ✔

Black-veined White *(Aporia crataegi)* ✔

Bright Copper *(Paralucia surifera)*

Brimstone *(Gonepteryx rhamni)*

Buckeye *(Precis coenia)* ✔

Cabbage White *(Pieris rapae)* ✔

California Sister *(Adelpha bredowii)*

California White Tip *(Anthocharis lanceolata)*

Camberwell Beauty *(Nymphalis antiopa)*

Checkered White *(Pieris protodice)*

Cloudless Sulphur *(Phoebis sennae)*

Colorado Hairstreak *(Hypaurotis crysalus)*

Common Imperial Blue *(Jalmenus evagoras)*

Common Sulphur *(Colias philodice)*

Death's-head Hawk moth *(Acherontia atropos)*

Diana Fritillary *(Speyeria diana)*

Dimorphic Bark Wing *(Epiphile adrasta)*

Dotted Checkerspot *(Poladryas minuta)*

Eastern Comma *(Polygonia comma)*

El Segundo Blue *(Euphilotes bernardino allyni)*

European Map *(Araschnia levana)*

European Orange-tip *(Anthocharis cardamines)*

Field Crescent *(Phyciodes pratensis)*

Fruit-feeding Charaxes *(Charaxes eupale)*

Glanville Fritillary *(Melitaea cinxia)*

Goldenrod Stowaway *(Cirrhophanus triangulifer)*

Grapevine Epimenis *(Psychomorpha epimenis)*

Gray Cracker *(Hamadryas februa)*

Grayling *(Hipparchia semele)*

Great Copper *(Lycaena xanthoides)*

Great-Spangled Fritillary *(Speyeria cybele)*

Great Southern White *(Ascia monuste)*

Green Hairstreak *(Callophrys affinis)*

Gulf Fritillary *(Agraulis vanillae)* ✓

Gypsy moth *(Lymantra dispar)* ✓

Heath Fritillary *(Mellicta athalia)*

Hornet moth *(Sesia bembeciformis)*

Large Blue *(Maculinea arion)*

Large Heath *(Coenonympha tullia)*

Large White *(Pieris brassicae)*

Long-tailed Blue *(Lampides boeticus)*

Long-tailed Skipper *(Urbanus proteus)*

Madrone *(Eucheira socialis)*

Marbled Fritillary *(Brenthis daphne)*

Marbled White *(Melanargia galathea)*

Meadow Brown *(Maniola jurtina)*

Monarch *(Danaus plexippus)* ✓

Moroccan Orange Tip *(Anthocharis belia)*

Mourning Cloak *(Nymphalis antiopa)*

Northern Blue *(Plebejus idas)*
Orange Sulphur *(Colias eurytheme)*
Owl *(Caligo atreus)*
Painted Lady *(Vanessa cardui)*   ✓   *Calif. Lady* ✓
Palos Verdes Blue *(Glaucopsyche lygdamus palosverdesensis)* ✓
Paradise Birdwing *(Ornithoptera paradisea)*
Peacock *(Inachis io)*
Pipevine Swallowtail *(Battus philenor)*
Postman *(Heliconius melpomene)*
Purple Emperor *(Apatura iris)*
Queen *(Danaus gilippus)*
Queen Alexandra's birdwing *(Ornithoptera alexandrae)*
Queen Swallowtail *(Papilio androgeus)*
Red Admiral *(Vanessa atalanta)*   ✓
Red Cracker *(Hamadryas amphinome)*
Red-spotted Purple *(Limenitis arthemis)*
Ringlet *(Aphantopus hyperantus)*
Scarlet Tiger moth *(Panaxia dominula)*
Sharp-veined White *(Pieris napi)*
Silver-spotted Skipper *(Epargyreus clarus)*
Silver-washed Fritillary *(Argynnis paphia)*
Small Postman *(Heliconius erato)*
Snout *(Libytheana carinenta)*
Sonora Blue *(Philotes sonorensis)*
Southern Dogface *(Colias cesonia)*

Southern Festoon *(Zerynthia polyxena)*
Speckled Wood *(Pararge aegeria)*
Spicebush Swallowtail *(Papilio troilus)*
Tiger Swallowtail *(Papilio glaucus)*
Tortoiseshell *(Nymphalis vaualbum)*
Two-tailed Pasha *(Charaxes jasius)*
Viceroy *(Limenitis archippus)*
Western Tiger Swallowtail *(Papilio rutulus)*
White Admiral *(Ladoga camilla)*
Wood White *(Leptidea sinapis)*
Yucca moth *(Tegeticule yuccasella)*
Zebra *(Colobura dirce)*
Zebra Longwing *(Heliconius charitonia)*

ONE

Obsession
with Butterflies

IN PHYSICS, STRING THEORY SUGGESTS THAT there are more than four dimensions, perhaps ten in all. These extra dimensions are curled up into a very small space, big enough only for subatomic particles, or tiny loops of vibrating "string." The theory does not rule out more dimensions, perhaps in the area of time. These dimensions, here but not here, exist outside our range of perception.

Adding butterflies to your life is like adding another dimension. The air trembles with the movement of wings. The approach of a White Admiral. The aerial dance of sulphurs. A Painted Lady. A Black Satyr. All this existed before, has always existed, but you were unaware. You didn't see. At various times and places, in winter, or on a busy street, the air is still and butterflies are impossible. Yet their presence remains, like one of those other ten dimensions. You've added this to your life.

Butterflies became present in my life one summer afternoon by a river in New Mexico. A Western Tiger Swallowtail dipped by my face. About three inches across, it seemed much larger. Its lemon yellow wings were striped improbably and fluted in black. They filliped into a long forked tail with spots of red and blue. Smelling nothing of interest, the butterfly floated away, leaving me pleased and agitated, as though I had been handed a gift I didn't deserve. Could this, all along, be a simple truth: beauty without cause or consequence?

The Western Tiger Swallowtail was patrolling for a mate, avoiding birds, and on the lookout for nectar or carrion juices. Like most butterflies, it tasted with its feet and smelled with its antennae. Its genitalia had eyes, simple light-sensitive cells. It had been alive for a day. It might live another month.

Later, I became enamored with the tiniest of butterflies, thumbnail-sized gray hairstreaks in my peripheral vision, on a weed or a fence, common as a mailbox. But wait until they settle and show their underside. Scallops of mango orange. Patterns of blue and russet. A crescent, a dash, a language in code.

In the second movie of the *Jurassic Park* series, actor Jeff Goldblum is once again trapped on an island filled with dinosaurs. As the other characters admire a herd of tricer-

Western Tiger Swallowtail

atops, Goldblum says dryly, "Ooooh! Ahhhh! That's how it always starts. But later there's screaming and running."

Oooh. Ahhh. That's how it starts. Later there are guide-books and more guidebooks and picnics in meadows and screaming and running. Some of us become obsessed with butterflies, although I would never include myself in that category. I am interested, yes, but not obsessed.

Not like those other people.

Eleanor Glanville was a woman of property and modest wealth, thirty-one years old, seven years a widow, the mother of two children. In 1685, she married again, a man ten years younger. This time, she married badly.

When her second husband cocked his pistol and pointed it at her breast, shouting that he would shoot her dead, did Eleanor think of Purple Emperors falling through the sunspots of an oak woodland? When the same man left her, after the birth of two more children, did she find peace in the rearing of her caterpillars, the Large Whites feeding on the leaves of watercress, cabbage, and turnip, the fritillary changing into its pupa or chrysalis, "a thick larg cofen ye same coler speckt with a row of silver on each sid"?

In 1703, a well-known entomologist in London wrote that Lady Glanville had come "to town with the noblest collection of butterflies, all English, which has sham'd us; her way is to give for forty or fifty ordinary caterpillars six-pence, and to feed them; if a fine caterpillar, for encouragement, six pence a piece, which is one way to employ the poor."

The lady had already sent cases of butterfly specimens to the premier naturalist of her day, James Petiver, who responded with gratitude and admiration. Included in the collection was the first record of the Glanville Fritillary, a prettily patterned flutter of orange that "breeds on steep and broken declivities near the coast, which the scythe or the plough never as yet have invaded."

By now, the evil second husband, Richard, had a new mistress and a new baby and was scheming to disinherit his first son by Eleanor. This seventeen-year-old boy was serving as an apprentice for James Petiver when his father kidnapped him, held him, and bullied him into renouncing his mother and his inheritance. Richard Glanville also worked to alienate Eleanor from her other children, so that at her death she left most of her property to a second cousin. Another of Eleanor's sons challenged the will, declaring that his mother had made it under the mistaken belief that her children had been turned into fairies.

As early as the Middle Ages, people believed that butterflies, or *buterfloeges*, were disguised fairies bent on stealing dairy products such as butter, milk, and cream. Over time, fairies and butterflies became ever more linked, both tiny winged creatures, sportive, and seemingly merry.

Eleanor Glanville may only have been hoping for the best.

During the contest of her will, one hundred witnesses came forward to testify. Her former neighbors were quick to remember her strange behavior: how she dressed like a gypsy, appeared on the downs "without all necessary cloathes," and "would carry a sheet out under the hedges and bushes and with a long pole beat the said hedges and cach't a parcel of wormes."

Friends such as Petiver and other scientists appeared in Lady Glanville's defense. Still, the verdict overturned her final wishes on the grounds of insanity. As one entomologist later admitted, "None but those deprived of their Senses would go in Pursuit of butterflyes."

That sentiment would change. In the mid-eighteenth century, English butterfly collectors began calling themselves Aurelians, from the Latin *aureolius,* a reference to the golden chrysalis of some species. These men and women might still be considered odd, armed as they were with oversized nets and satchels of paraphernalia, but they were more mocked than scorned, viewed even with affection.

Historian David Allan believes that "the eighteenth century was a period of transition. In its earlier years we can watch people playing with nature, treating it like a newly purchased toy. Later, as they become accustomed to the novelty and learn to react with less and less unease, we see their boldness grow. Eventually, as the century ends, we find them helplessly in love with it."

By the Victorian age, in the mid-1800s, nature had become part of the household furniture, represented by the curio cabinet filled with minerals, fossils, dried plants, and seashells. The impulse mixed science with a collector's greed.

And butterflies, so bright, so distinctly patterned, were eminently collectible. Seemingly, every other good man, sometimes his good wife, and often his unruly children were obsessed with the insect. Lectures, social clubs, and field trips welcomed people of all classes, and people of all classes came: to learn the habits of a Meadow Brown, to catch a Camberwell Beauty, to revel in the abundance of Silver-washed Fritillaries dancing over sweet-scented bramble.

Like fairies, this abundance is something we can only imagine. At that time, there were miles of meadows, pastures, hedges, and woods; no cars, no chemicals, millions fewer people, and thousands more coppers, blues, whites, and sulphurs swirling like confetti in the air. The stout-hearted members of the Berwickshire Naturalists Field Club or the Haggerstone Entomological Society could hardly guess then what kind of party they were celebrating, or how it would end.

In 1876, Walter Rothschild, an eight-year-old boy in a wealthy banking family, started his own natural history museum and hired a skilled taxidermist as his first assistant. By the time of his death sixty-three years later, Lord

Rothschild was recognized as the world's greatest butterfly enthusiast, an eccentric who harnessed zebras to his carriage, which he rode through Piccadilly to Buckingham Palace, a statesman who helped forge the 1917 promise of a homeland for Jews in Palestine, a collector who bequeathed his set of 2.25 million butterflies and moths to the British Museum in London, making it the holder of more lepidoptera anywhere at any time.

Lord Rothschild did not collect 2.25 million butterflies by himself. He hired professionals, men, and later women, in the business of going to remote places. One of these was the Australian collector, A. S. Meek, who traveled mainly through Papua New Guinea and the Solomon Islands. Meek sent back thousands of new species, including the largest of all butterflies, the Queen Alexandra's Birdwing, the female nearly a foot long, the male iridescent green and luminous blue, with a bright yellow abdomen.

The biological diversity of New Guinea lies in its life zones, which range from hot lowland forests to snowy peaks. During one trip in the cold mountains, most of Meek's bearers fell ill. Meek himself was nearly giddy with the recent capture of a new female birdwing, a high-altitude specialist with a hairy body.

He notes that it was particularly vexing to be faced with

making the decision either to stay where the collecting was so rich or to return to the coast to save his men. One by one, the natives contracted pneumonia. Their breathing grew labored, and they seemed close to death.

Often enough, Meek had been in the same situation, ill and shivering with fever. "I suppose," he wrote, "that the people of civilized countries will wonder that there was any doubt for a single moment in my mind, as to whether the health—and perhaps lives—of the boys should be sacrificed for the sake of collecting a few butterflies. But in the wild world, away from the ideas of civilization, one gets what I would not call a recklessness or an indifference to human life so much as a somewhat different idea of its value. A certain work to be done seems to be a bigger consideration."

Eventually, a young man dies, and Meek returns to the coast.

Similar scenes were being played out around the world. Men faced danger and disease, a collecting net in one hand, a gun in the other. (More than one collector used that gun to shoot down the luminous birdwing as it flew out of reach at the tops of trees.)

In the United States, in 1871, Theodore Mead headed west on a butterfly-collecting trip and wrote home laconically:

There are several hotels here in Denver. The one where we are staying is good but rather dear for a country village (4.50 a day). We start on Monday morning to Fair Play in South Park, by stage, 17 hours. Indians are friendly—they only killed one man last week, 12 miles from Greely. . . . As there are no large bands in South Park, I don't think we run very great risks. I met a former acquaintance here, now territorial officer in Wyoming. He did not like our idea of camping out alone and said we might go 40 times without accident but the 41st time they would "gather us in."

Another explorer in Colorado had collected a brilliant group of new specimens until one night when two of the men driving his pack train stole his supplies, fished out the insects, and drank the refreshing alcohol used to preserve them.

By now, many collectors were also naturalists. In 1898, the author of a guide to the butterflies of the eastern United States could write with surprising accuracy about the shape of butterfly eggs, the habits of caterpillars, and the physiology of adults, including the scent scales on the wings of a Mountain Silver-spot and the structure of a swallowtail's antenna. The life history of a species became as important as its naming and death at the end of a pin. The twentieth century saw more and more people following the flight of a butterfly, not to capture it but to see where it went and what it did with its time.

There are some 18,000 species of known butterflies and 147,000 species of moths, both in the order Lepidoptera. Briefly, the differences between a butterfly and a moth are that most butterflies fly during the day, most moths do not; most butterflies have bright colors, most moths do not; most butterflies have distinctly clubbed antennae, most moths do not; most butterflies rest with their wings clapped above their bodies, most moths do not. At the same time, most moths have rather hairy bodies, and most butterflies do not; most moths have hooks linking their forewings and hindwings, and most butterflies do not.

Of what use are butterflies? Less than you might think. Butterflies are not like beetles or bees, engines of pollination. They do not even compare favorably to moths, their kissing cousins. If all butterflies were to disappear, so would a few flowers—but not many. (If all flowers were to disappear, so would we, since almost everything we eat depends upon a flowering plant.)

The Taoist master Chuang Tze, for whom uselessness had a certain grace, and even a certain use, wrote: "I dreamed I was a butterfly, fluttering hither and thither. I was conscious only of following my fancies as a butterfly, and was unconscious of my individuality as a man. Suddenly I awoke and there I lay myself again. Now I do not know whether I was then a man dreaming I was a butterfly or whether I am now a butterfly dreaming I am a man."

A modern interpreter of Chuang Tze reminds us that characterizing "life and knowledge as dream is not to denigrate its reality." Dream does not imply delusion but a "radical interchange among separate identities."

To exchange identity with a butterfly is radical. It is to be what you are obviously not. It is to find surprising connections to the world, as well, perhaps, as hidden dimensions, small but powerful, outside your range of perception.

Moreover, the life of a butterfly is the enactment of myth. As "a parcel of wormes," caterpillars crawl lowly on the ground. They hide in the debris of leaves and twigs. Some of these larvae bristle with spines to deter predators. They are as crudely colored as a child's wooden toy. They spit acrid vomit and emit poisonous gas. From this dubious state, they form their hard, protective chrysalides and enter a sleep in which they transform themselves.

The adult emerges. It rises like a phoenix.

And we, who live by myth, who live in fear of change and in fear of death, are privileged to see this metamorphosis over and over, a common thing, an everyday thing for a fat green "worme," a bag of goo splotched with yellow, to transform into a Western Tiger Swallowtail, fluted and glowing.

The French naturalist Marcel Roland said that butterflies give us "solace for the pain of living."

I would venture to guess that more people today are obsessed with butterflies, looking for solace, than ever before. Many are graduate students and professors. Some use butterflies as models to examine issues of genetics and insect biology; these scientists dutifully apply their research to agriculture and conservation. But most study butterflies for less practical reasons, often for the simplest of human motives.

Miriam Rothschild, born in 1908, was the niece of Lord Walter Rothschild and the daughter of Charles Rothschild, a man who once stopped a train when he saw a rare butterfly through the window. Charles Rothschild reserved his finest passion for fleas, however, and his daughter went on to produce a six-volume inventory of his collection of several million, claiming for herself the title of "Flea Lady."

In her research on butterflies, Miriam showed how Monarch caterpillars ingest and store the poisons in milkweed plants. She looked closely at the role of pigmentation in a butterfly's chrysalis or pupa. She demonstrated that Large White females use chemical cues to avoid laying eggs on leaves that already have eggs or feeding larvae.

These butterflies want the best for their young: an abundant food source, without competition.

During the twentieth century, Miriam Rothschild helped extend the work of the nineteenth-century naturalist and collector, men like her father and uncle, into the world of ecology and biochemistry, of molecules and odor plumes and secret signals. The Large White had been pinned, named, dissected, and observed in the field. But there was more to do.

Musing over why the pupa of a Large White is sometimes blue, its bile pigments remaining in the surface tissues, Miriam Rothschild asked, seemingly without guile, "Who will elucidate this mystery?"

Who will elucidate the mystery of the morpho caterpillar, which exudes a drop of clear liquid and studiously combs it through all its tufts of hair?

Who will elucidate the mystery of why some butterflies court and some rape?

Of how Postman butterflies remember and avoid the spot where a researcher once netted them?

Of butterflies with ears on their wings?

Of number, and extinction, and of how many butterflies exist in the world?

There is more to do, and that is true still for places such as Papua New Guinea and the Solomon Islands, where

John Tennent is a scientific associate for the British Natural History Museum. During a recent collecting trip, he came across a flowering tree on the island of San Cristobal. In brief, intense periods of producing nectar, such trees attract a variety of butterflies. John caught three with one swoop of his net, a male and two females of two unknown species. Over the next few days, he collected a "fair series of one" but not of the other, although he visited the tree regularly, then and months later. "That species was not seen before," John says, "and has never been seen since."

A small blue butterfly now bears the name *Psychonotis julie,* after John's wife.

On his last trip, in the first year of the twenty-first century, John was marooned on the island of Tikopia for eight weeks due to a local coup d'etat. Unfortunately, Tikopia has only thirteen species of already discovered butterflies. John passed the time killing flies for the lizard that lived on his doorstep and teaching the island children interminable verses of "Old MacDonald Had a Farm."

He is now ready to publish his book on the biogeography of the Solomon Islands, which will include descriptions of seventy new butterfly species and some interesting observations on mimicry.

"Butterflies add another dimension to the garden," Miriam Rothschild wrote, "for they are like dream flowers—childhood dreams—which have broken loose from their stalks and escaped into the sunshine. Air and angels. This is the way I look upon their presence, not as a professional entomologist."

There comes a moment in your life when you must look at what you love and think: Yes, I was right.

People who love butterflies have it easy.

TWO

TOUGH LOVE

$\mathcal{A}$ FEMALE BUTTERFLY LAYS AN EGG THAT looks like a miniature pearl, or a squashed golf ball, or a whiskey barrel. She might lay one egg or a clutch of many.

The danger begins at once. Viral, bacterial, and fungal infections can attack the egg. Tiny parasitic wasps or flies burrow into its tissue and lay their own eggs; when these young hatch, they feed on the embryonic caterpillar. In the adult female Owl butterfly, parasitic wasps ride on the mother's hindwing and jump off like pirates as she deposits her treasure. An assassin bug passes by and eats the clutch for breakfast. A deer eats the leaf on which the eggs are laid. The possibility for disaster is high; the chances for survival are not.

If the butterfly is a skipper, the emerging caterpillar will start by eating its eggshell and then rear up grandly like a cobra from its basket. If the butterfly is a swallowtail, the larva will look pathetically damp and fragile, its hairs pressed flat to its body.

A young caterpillar can be the size of a comma or a hyphen or a dash. Already it has its major parts: a hardened head with mandibles for biting and chewing food, three thoracic segments, each with a pair of jointed legs, and ten abdominal segments with five pairs of "false legs," or small hooks. Pores in the skin open and close to let in air. The whole thing ends in an anal plate, which can confusingly resemble the front.

This hyphen is aware of its world. Along each side of the head are simple eyes with photosensitive pigments. Swaying from side to side, the larva sees a mosaic of the visual field, easily telling light from dark, horizontal from vertical. Odor detectors, or noses, are scattered the length of the body: on the antennae, on the abdomen, on the legs. Some of the olfactory hairs double as taste buds. Other hairs perceive touch. Some detect sound or vibration.

Above and between the caterpillar's mouth parts is a tube that produces silk. Most larvae spin out a sticky thread that helps secure them to surfaces as they move forward. In this way, they are not easily shaken from a leaf or twig. Silk can also be used for rolling up leaves, creating shelters, attaching the pupa to its resting spot, or spinning the rare cocoon.

The California White Tip lays an egg that resembles a tiny ear of corn. Miraculously, the egg hatches. Without pause, the caterpillar begins to feed and grow until the joints between its body segments distend. This distension activates hormones. A new exoskeleton forms inside the old, which will be partially dissolved by enzymes. The caterpillar rests, takes in a "gulp" of air, puffs out its segments, and the old skin splits at designated seams. The larva has molted into its next stage.

Each stage is called an instar, and, like explorers in time and space, caterpillars move from instar to instar, usually five. These stages are referred to as a first instar, a second instar, a third instar, a fourth instar, and a fifth instar—the graduating class.

Obsessed with food, caterpillars are a mouth attached to a stomach. The cliché of the caterpillar is that it is a "voracious eating machine." Over and over, in books and articles, entomologists repeat this phrase with admiration and envy.

Eating, growing, resting, molting, eating, growing, resting, molting, larvae must obtain all the nutrients they need to develop into a fully grown butterfly. This often includes the proteins necessary to produce eggs and sperm. Some moth species gain over 3,000 times their hatching weight. In human terms, a 10-pound baby becomes a 30,000-pound man.

The appearance of caterpillars varies enormously. It's Halloween night, and a costume ball. In different species, the skin that encloses the stomach may be smooth or bumpy, covered with hairs or spines, erupting in filaments or horns. The shape of the caterpillar may be long or short, slender or fat. Some caterpillars look like slugs, some like brown twigs. A few seem to be preparing for Mardi Gras with an array of appendages like a headdress of balloons.

The design can be as gaudy as a finger painting or as sophisticated as an Escher print.

The Cabbage White is a minimalist, bluish green with a simple yellow line. A Buckeye is bristly and black with two rows of orange spots and two of creamy yellow; the back spines have blue bases; the side spines have orange. The American Painted Lady has been described as a "truly beautiful caterpillar" with yellow-green stripes and red and white spots on black bands. Decorating a daisy, the Lady retreats into her nest of silk like a starlet firmly shutting the door.

If caterpillars are obsessed with food, a number of animals are obsessed with caterpillars. The butterflies we see flying through the garden are extraordinary not only for their beauty but for their survival. Nearly every one of their siblings has gone into the food mill of nature: infected by a pathogen, parasitized by a wasp, or eaten by a bird.

Caterpillars need luck and strategy.

As an individual molts, changes, grows bigger, and becomes more obvious to predators, it needs better luck and new strategies.

The first instar of a Western Tiger Swallowtail is tiny, mottled, hairy, a mere speck on a leaf, hard to distinguish from a bit of dirt.

The second and third instars look like bird droppings, a saddle of slick white on brown. This is a droll trick. Birds don't eat their own droppings.

The fourth instar of a Western Tiger Swallowtail is green and smooth with orange-yellow eyespots that have a blue center. The green helps the caterpillar blend into its background. The eyespot might scare off a small bird.

The fifth instar has grown large enough to resemble the front end of a bright green snake with two colorful, conspicuous eyespots and a wide, wide mouth. (Some researchers think this larva is trying to resemble a bad-tasting tree frog, not a snake.)

Typically, in most species, later instars are hairier, spinier, bristlier, and meaner-looking. They may have new fleshy filaments. They may look like a walking hairbrush. If handled, those hairs might leave you with an itch or a rash or a sting. The message is getting clearer: I'm not worth eating.

As they move into their fourth or fifth instar, caterpillars may need to change their behavior as well as their appearance. They may begin feeding at night instead of in the day. They may eat differently.

Birds spend a lot of time looking at plants, and birds recognize vegetation that has been damaged by caterpillars. Black-capped chickadees are adept at spotting bitten, tattered leaves, as well as the guilty insect hiding nearby. Blue jays can tell the difference between photographs of whole leaves and those of partially eaten leaves.

Caterpillars respond by paring leaves along the side, making large leaves look like smaller, intact leaves.

Now the blue jay is confused. Is that an undamaged leaf, a leaf pared by a caterpillar, or another photograph?

Some larvae go so far as to eat holes in leaves and fill in the empty spaces with their bodies; these caterpillars may have markings that resemble leaf blemishes or a drying brown edge. One caterpillar has serrations along its back that mimic the serrations on the elm leaf.

When a leaf becomes too tattered, and the caterpillar's presence too obvious, the caterpillar might climb to a safe spot and snip the evidence from the tree.

Problem solved.

Caterpillars that are poisonous or taste bad to a bird don't need to work so hard. The Monarch is a messy eater.

Its coloring—stripes of yellow, black, and white—warns predators away. Other toxic, colorful species are also free to create "ghost plants," leaves munched to a skeleton of stems and veins.

In the American West, there's an old, sad song with the line, "You can see by my getup that I am a cowboy."

You can see by a caterpillar's getup that it is a green leaf or a brown twig or a bit of excrement or a tree frog or something that will make you sick.

How I see a caterpillar, of course, is not how a bird sees it. Chickadees have better vision than humans, seeing more colors (including ultraviolet) and seeing them differently. Distance, background, and shifting patterns of light all affect how well a larva is hidden. Even the striped Monarch may blend into its background when feeding on the undersurface of leaves.

For a caterpillar, it never ends. There is always one more thing to worry about. The mimicry or camouflage that works so well against a bird may not work at all against the predatory stinkbug, which has been known to stalk its prey for as long as an hour. Some caterpillars do the obvious. They drop off the leaf and hope for a soft landing. Or they spin out a thread of silk, drop like Tom Cruise in *Mission Impossible,* and dangle from the lifeline while they wait for the predator to leave.

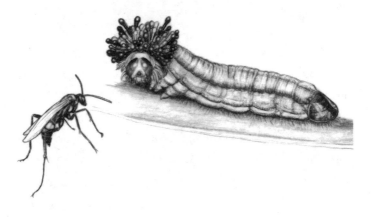

Wasp attacking Ms. Balloonhead

Some parasitic wasps wait, too, for their prey to climb back up. Some wasps slowly walk down the silken line. Some wasps slowly reel in that line . . .

In a battle that can be an even match, a caterpillar hunches and rears, increasing in size, trying to look as formidable as possible. It may wag back and forth and then lunge at its attacker in an effort to knock him down. Against wasps or ants, some larvae vomit a toxic green fluid. The social caterpillars of the Red Cracker emit a foul odor. The Western Tiger Swallowtail larva has an orange, forked, fleshy scent gland that pops out of its head. Its sudden appearance is alarming enough. Then the gland releases acid.

Occasionally, a caterpillar tries to make a run for it. Caterpillars move in waves of contractions that go from the tail to the head. Each segment rises from the ground, pushes forward into its neighbor, relaxes, and falls back to the ground. That's one step. A normal walking pace is less than half an inch per second. But if necessary, some larvae can pick up speed by reverse galloping. Now the wave begins at the head and the whole body arches up, wrenching the legs free of the surface. When the segments relax back to the ground, claspers on the tail detach and reattach and move a step back. Sometimes the raised body flexes sideways. Sometimes it coils into a wheel and rolls back at a speed of fifteen inches a second! This is a sprint, the ultimate chase scene.

In one experiment, researchers watched 628 interactions between wasps and Asian swallowtail caterpillars. Inexperienced and new to hunting, 178 wasps did not attack at all. Against the 450 who did, most caterpillars hunched up; extruded their orange, forked, fleshy scent glands; and wafted a nasty chemical. Deterred, 191 wasps backed away; 64 of these successfully reattacked. Twenty-six caterpillars chose to drop off the tree; nine of those survived. In all, the wasps caught half their prey.

In a dangerous world, hiding can be the better part of

valor. Most caterpillars hide in plain sight, through camouflage or mimicry. Some retreat as a group into nests made of silk. A skipper caterpillar builds individual leaf shelters, each instar producing a specific construction. First and second instars make two semiparallel cuts from the edge of the leaf, bring the flap over like a hinge, and secure the structure with silken guylines. Third instars make one cut from the leaf's edge and roll over a bigger part of the leaf to be tied down. Fourth and fifth instars either fold and secure a large portion of the leaf toward its center or tie two leaves together to make a pocket. Then they go inside their little houses, hoping no one will ever come to visit. (Skippers are known for forcibly ejecting their waste, called frass, at great speed, flinging it from their leaf houses for as much as five feet, and as fast as four feet per second. This prevents disease, smell, and an unsightly pileup.)

If there were a villain in this story—which, of course, there is not—it would be the parasitic wasp. Finding its host through odor cues, this wasp lays its eggs in the caterpillar's body. When the eggs hatch, the wasp larvae use the caterpillar as a food source, devouring it from within. In some species, many wasps emerge from their dead or dying host; in others, only one wasp develops.

Parasitic flies do something similar, laying their eggs directly on the caterpillar or sprinkling leaves with "micro-eggs," which the caterpillar eats. Micro-eggs are exactly what mothers are thinking about when they tell their children not to lick the grass.

Few caterpillars have enough strategies to outwit all their predators. The Baltimore Checkerspot is a black, white, and orange-tipped butterfly whose larvae live socially in silk webs. Their colors label them as poisonous to birds. Against other attackers, they throw up, jerk their heads, or escape into their nest. Despite these efforts, parasitic wasps often lay eggs in the early instars. The caterpillars hibernate or overwinter in their fourth instar, and so do the immature parasitoids. When the Checkerspot caterpillar begins to move about in the spring, an average of seven or eight wasp larvae burrow through their host's skin and spin cocoons directly underneath, tying the still-living prey to the vegetation. A week later, the mature wasps emerge and attack the instar.

To be served up as food is the main ecological role of the caterpillar. A Baltimore Checkerspot or Monarch or Large White lays hundreds of eggs. Only a few grow up to lay more eggs. The rest are meant to be found and enjoyed.

It gets worse.

The very plants caterpillars live on, the very leaves they are eating, are out to get them, too.

Actually, this seems fair. In any given forest, caterpillars are probably consuming more vegetation than all the other insects combined. Plants defend themselves directly with thick skins, sharp spines, wooly hairs, resinous or gluey secretions, and sawtooth leaf edges.

In one passion vine plant, the larger instars of a caterpillar are caught and held on small hooks. The scene is medieval. More subtly, the individual leaves of a passion vine can be so irregular that female butterflies have trouble recognizing the leaf shape on which to lay their eggs.

Passion vines may also have bumps that resemble eggs, discouraging adult butterflies from laying more. (Too many eggs in a caterpillar's lifestyle encourage cannibalism; the first instars hatch and begin eating whatever they can find, including nearby eggs and other first instars.)

In some species, plant projections actively promote egg laying on a site that the plant later discards, effectively getting rid of future caterpillars. Some leaves automatically die around the egg of a butterfly and the dead tissue drops away.

Other plants have pockets of toxic material scattered inside and throughout their leaves; each bite further poisons the caterpillar. Plants might also have canals or veins that pour out poisonous glue, trapping and killing the insect.

Such a lethal sap flows through the canal system of

milkweed plants. This works fine against most insects, except for the Monarch and its relatives who have adapted by taking in these chemicals and using them to become distasteful to predators. Miriam Rothschild once observed Monarch larvae lapping up the latex oozing from a vein like a "cat drinking milk." Another researcher saw Monarchs trying to limit their intake by pinching the veins of the leaf with their mandibles, damaging the canal system internally.

Many species cut a trench across a leaf blade and let the canals drain before they start feeding.

Architecture is a plant's first defense. Once a caterpillar begins to eat, compounds in its saliva can also be recognized by the leaf as a form of attack. *Whang, whang, whang.* The entire plant goes on alert. A hormonal burst starts a second defense system, a counterattack that may include rushing newly produced toxins to the damaged leaf, as well as compounds that slow the caterpillar's ability to digest the plant as food.

Making these compounds uses up expensive resources. And insects may adapt to the new poison. As a third defense, some plants also send chemical signals into the air, crying out unashamedly, "Help!"

The parasitic wasp hears the call. Other predators, such as the small but rapacious true-bug, hear the call. They follow the plant's odor plume, find the caterpillar, and attack it. From the plant's point of view, they are the cavalry.

Wounded plants also emit chemicals that tell moth species to avoid laying more eggs on the plant. The moth obeys because she knows that the plant's defenses have been activated by larvae who would compete with her own. Presumably, butterflies get a similar message.

This is that other dimension, here but not here, outside what we know: molecular cries for help, the scent of intrigue, secrets in the air.

Layers upon layers. The chemical mix that a plant sends out can signal which species of caterpillar is feeding on the leaf, telling choosy parasitic wasps that this caterpillar will be the right host for their young.

How can a plant distinguish between different insect species or between a caterpillar bite and some other kind of wound? Researchers think that bacteria living in the caterpillar's gut may produce the compound.

It's just another betrayal, in a Dickensonian childhood.

THREE

You Need a Friend

*I*N A DANGEROUS WORLD, IT IS GOOD TO HAVE friends. Even caterpillars have a few. Most of them are ants, but I also count people such as Philip DeVries, a MacArthur Fellow who has been working with caterpillars for thirty years.

As an undergraduate, Phil studied botany. A lot of his friends studied butterflies. When these men went out in the field, they raced around with nets and voices raised in triumph. Phil couldn't compete. So he changed the rules: He found what the caterpillars were eating. Then he found the caterpillars.

"How satisfying," he would write later, slyly, in his book *The Butterflies of Costa Rica*, "to observe a butterfly and be able to associate it with a host plant and a series of interactions within a habitat; how indolent to simply swing the net and collect a specimen."

Phil's research into the association of ants and caterpil-

lars began in Borneo when he watched a group of ants ne-
gotiate with the larva of a blue butterfly. He could as easily
have been in Canada or Central America or England. Over
2,000 species of butterflies around the world suffer from
myrmecophily. They are ant-loving.

"This was a subject," Phil says, "that I knew would keep
me interested for a long time."

In Panama, Phil experimented with metalmark caterpillars who feed on sapling trees that have a small nectary at
the base of each leaf. These nectaries produce a sweet liquid that attracts ants who guard the plant against other destructive insects. These same ants attend the metalmarks
and protect them against predatory social wasps.

Because this is what biologists do, Phil placed caterpillars on plants with ants and on plants without ants and
timed how long the larvae would survive. If ants were present, they vigorously defended the caterpillar. If ants were
not present, a wasp zoomed to the leaf, stung the larva, cut
up the body, and carried the parts away. Within minutes,
unprotected caterpillars were being fed to hungry wasp
grubs.

The wasp's behavior seemed more reasonable than the
ant's.

But ants have their own logic. They ignore young metalmarks and wait for the larva's third instar, when it has a

number of new organs. Now when an ant strokes the caterpillar's back, a pair of glands emerge, looking, Phil says, like the fingers of a rubber surgical glove. The glands secrete a drop of clear fluid that the ants drink with seeming pleasure.

Ants are so eager for this honeydew they will stroke the caterpillar over and over again. Phil estimates that ants attending a metalmark solicit their new friend at least once a minute. When it grows tired of the attention, the caterpillar audibly taps the ground. Like scolded children, the ants stop—for a while.

Instead of returning to its nest, an ant may stay with a metalmark for over a week, kept there by a series of tricks and a false sense of duty.

At the front of the caterpillar, two tentacle organs seem to emit a chemical alarm similar to the one ants use when they signal each other. When these organs emerge, Phil says, he has seen attending ants "snap into a defensive posture, mandibles agape, abdomen curled under the body." When he mischievously moves a bit of twig or grass near the caterpillar, the ants rush at the object in a fury, stinging and biting it.

At the top of the caterpillar's head, two more rodlike organs called papillae produce a sound, or vibratory call. A former jazz musician, Phil compares the caterpillar's papillae to the guiro, a Latin American instrument used in salsa or bossa nova music. The guiro is played by sliding a wooden stick across the grooves in a carved gourd. Similarly, when the caterpillar moves its head, the ringed shaft of each papilla grates on specialized bumps, or ridges. This vibrational song mimics the call of ants trying to attract each other's attention.

For the metalmark caterpillar, a partylike atmosphere of good food, good music, and chemical signals, or pheromones, guarantees a supply of willing bodyguards. As a bonus, the caterpillar gets to drink from the plant's nectaries, what the ants were originally protecting.

Elsewhere, in Australia, the Bright Copper caterpillar is spending its day in an elaborate underground chamber constructed and maintained by ants. These chambers can hold up to twenty butterfly larvae and ten chrysalides, or pupae. At sunset, the caterpillar emerges to feed on leaves; at sunrise, it descends again into the shelter. Day and night, a Bright Copper larva may be attended by as many as twenty-five ants. In return, both larvae and pupae provide a sweet liquid, a nice mix of glucose and amino acids.

The caterpillars of the Australian Common Imperial Blue butterfly have three calls to which ants respond: a rapid hiss-hiss-hiss, a kind of grunt, and a low-pitched drumming. The hisses occur in the first five minutes after the larva has been discovered by a worker ant. The grunts are produced later while the ants are in attendance. The drumming is made by both tended and untended larvae.

In Asia, a carnivorous blue larva eats aphids, which are also cared for by ants, who also care for the caterpillar. An ant's preference seems to be for the honey of the blue larva rather than the honey of the aphids. Thus, the latter are cheerfully sacrificed.

In Europe, another blue larva resembles a monstrous ant grub. Ants find and carry these larvae to their nests where the caterpillars live for months eating the ants' real grubs, which the parents generously provide. Describing this scene, writer and lepidopterist Vladimir Nabokov exclaimed in disgust: "It was as if cows gave us Chartreuse and we gave them our infants to eat!" Although the caterpillars secrete honeydew, their sum effect is to harm the nest. They are, in fact, parasites who may have evolved from a relationship once mutually beneficial.

The tragicomic relationship of the English Large Blue and the wood ant was discovered as early as the 1920s. Soon after, in a BBC interview, novelist Sir Compton McKenzie imagined the caterpillar as Persephone carried off to the Underworld by a merry party of intoxicated ants. There, Persephone eats ant grubs, pupates, and emerges. Now her true nature is revealed. Her honey glands are gone, and her

Ant caring for third instar caterpillar

hosts are angry. She escapes her pursuers by exuding a sticky substance that entangles their feet. Quickly she makes her way to the surface; there, under the broad blue sky, she spreads her wings "and flies off like a minute bit of heaven. The eggs are laid again on the banks of wild thyme, and in due course the caterpillar . . . takes to meat."

In 1979, England's Large Blue butterfly became extinct when the rabbits who ate the long grass in the West Country died from disease. With the rabbits gone, long grass outgrew the short grass preferred by wood ants and hid those patches of thyme where Large Blue females laid their eggs. A different species of ant now dominated the area. When *they* found a Large Blue caterpillar, they ate it.

Later, a closely related, or possibly identical, Large Blue from Sweden was reintroduced and carefully nurtured, along with the short grass, the thyme, and the rabbits.

"It's so complex. It's so intertwined," Phil DeVries says. Because he studies tropical butterflies, and because tropical butterflies seem to be disappearing along with their rainforests, Phil complains that "he writes epitaphs for a living."

"It's pathetic," he continues, "how few life histories we really know."

Caterpillars are generally loners. Perhaps 10 percent of species associate with ants. Another small percentage could be described as "social" and associate with each other.

Social larvae, who congregate and feed together, are often playing a game of odds. There is protection in numbers: Individuals on the outer edge may more easily get parasitized, but if you are a core member, the chances are higher that you will not.

Perhaps, too, some young larvae cannot individually attack tough leaves but are better able to feed as a group. These caterpillars are often warningly colored, signaling en masse that they taste bad.

In Mexico, the Madrone caterpillar lives in colonies at elevations of 8,000 feet. The Aztecs called them *xiquip-ilchiuhpapalotl* (butterfly that makes a pouch) because of their brilliant white nests of silk, twenty or thirty in a tree, which could house several hundred siblings. The *xiquip-ilchiuhpapalotl* leave the nest to feed during the night and use it for warmth during the day. Males outnumber females four to one and do most of the construction and maintenance. They also create trails by laying out silk paths to foraging sites, leaving chemical cues for other caterpillars to follow. Commonly, the "altruistic" males die of malnutrition and exhaustion. The females grow twice as big as the males. They conserve their energy and consume as much as they need to produce eggs as adults. In spring, when it is time to form their chrysalides, Madrones pack themselves close together.

The communal urge to pupate is unusual. Most caterpillars not only prefer to be alone, they also feel the need to leave their host-plant in search of a place more protected and less obvious, away from all these damaged leaves. (Skipper larvae are an exception, curling up inside their leaf nests.)

How long does the party go on? How long does a caterpillar live?

Because madrone leaves have little nitrogen, a *xiquip-*

ilchiuhpapalotl requires eight months to get all the protein it needs before pupation. The rare carnivorous caterpillar may require only three weeks. Flower- and fruit-eating caterpillars consume enough food in four weeks. The larva who eats leaves could need eight. On less nutritious grass, a caterpillar might take three months; on hard-to-digest roots, it may be twice that. In very cold climates, with short growing seasons, the larval stage will last two to three years.

How long you live as a caterpillar can also depend on how long you live as an adult butterfly. If you are a bad-tasting butterfly, with a good defense against predators, you might want to shorten your time as a larva, when you are more susceptible to parasites. This means that you enter your adult life with fewer reserves and so you must seek out more nitrogen-rich nectar. You might even, like the Postman and Zebra Longwing, evolve to eat pollen.

If you are more vulnerable as an adult than as a larva, you might prefer to extend your childhood, using it to gather the nutrients necessary to mate quickly and repro-duce as a butterfly.

At some point, the bell rings. The hour is nigh. Hor-mones signaled by growth in your body joints have con-trolled the process of each molt. In the fifth instar, the last molt has stretched the joints between your segments for the last time. Now the production of juvenile hormones stops. A new directive begins flipping on switches in the genes of certain cells.

There is some thought, too, that the yellow carotenoid pigments in your blood are sensitive to light energy and may help you count time, telling some species that the days are growing shorter, that you must soon find that protected place.

The larva of the Large White can distinguish between fourteen and a half and fifteen hours of light. When internal conditions are ready, and when there are more than fifteen hours of light a day, the caterpillar will become a chrysalis that produces a Large White butterfly in less than two weeks. When internal conditions are ready, but when there are fewer than fifteen hours of light a day, the caterpillar will become a chrysalis that waits through the winter.

The blood of the caterpillar is counting time.

You are biding your time until, after two weeks or two months or two years of life, you change color, void your gut, and begin a brief stage of wandering.

You start to mosey.

You amble onto a front porch.

You crawl restlessly across the path.

Something important is about to happen.

FOUR

METAMORPHOSIS

$\mathcal{V}$LADIMIR NABOKOV MAY BE THE MOST FA-
mous lepidopterist of the twentieth century. Many people
know him for his novel *Lolita;* college English majors have
a broader appreciation of his work; entomologists still talk
about his reclassification of North and South American
blues. Nabokov wrote twenty-two scientific papers, discovered a few species, and worked for six years as a fellow at
the Harvard Museum of Comparative Zoology. Some of his
research on butterflies was seminal; but his greater legacy is
how he wrote about butterflies, how he conveyed his passion, which he called his demon, to the nonpassionate, be-
mused, but still willing-to-be entertained world.

"It is astounding how little the ordinary person notices
butterflies," Nabokov marveled, and he set about moving
that mountain, not out of altruism but because he had no
choice: He was a writer who noticed butterflies all the
time.

In a lecture he delivered in the 1950s at Cornell University, by way of discussing Franz Kafka's *Metamorphosis* and Robert Stevenson's *Dr. Jekyll and Mr. Hyde*, Nabokov was sidetracked by the transformation of a caterpillar. His sympathies were with the larva's growing discomfort, that "tight feeling" about the neck, the imminence of public implosion and disgrace.

"Well," Nabokov began,

the caterpillar must do something about this horrible feeling. He walks about looking for a suitable place. He finds it. He crawls up a wall or a tree trunk. He makes for himself a little pad of silk on the underside of that perch. He hangs himself by the tip of his tail or last legs, from the silk patch, so as to dangle head downward in the position of an inverted question mark and there *is* a *question*—how to get rid now of his skin.

Nabokov described the condition of the prepupal caterpillar, hanging upside down for hours at a time before it makes the final bid for pupation. At last, there is a wiggle, a working of the "shoulders and hips."

"Then comes the critical moment . . . the problem now is to shed the whole skin—even the skin of those last legs by which we hang—but how to accomplish this without falling?" The professor paused. We can imagine his literature students, puzzled, not quite transfixed.

"So, what does he do," Nabokov repeated,

this courageous and stubborn little animal who is already partly disrobed? Very carefully he starts working out his hind legs, dislodging them from the patch of silk from which he is dangling, head down—and then with an admirable twist and jerk he sort of jumps *off* the silk pad, sheds the last shred of hose, and immediately, in the process of the same jerk-and-twist-jump he attaches himself anew by means of a hook that was under the shed skin on the tip of his body. Now all the skin has come off, thank God, and the bared surface, hard and glistening, is the pupa.

Other entomologists have noticed this admirable twist (mainly used in two families, the brush-footed and snout butterflies): that moment as the crowd sits silent and the acrobat hangs suspended from swing to swing. Then the hook locks securely into place.

There is no safety net, which may be why most caterpillars, among them swallowtails, whites, and sulphurs, first spin a silk girdle that connects them to the surface of the wall or twig. Some larvae dangle from their heads, not their tails. Skippers sew together a shelter of leaves. Apollos make a loose cocoon.

The skin has come off, and the bared surface, the hardened chrysalis, or pupa, underneath is a vague, irregular oval. It may have hairs, horns, spines, or honey glands to feed friendly ants. Some features of the developing butterfly inside the chrysalis can be recognized in its shape: the wingpads, the curve of the thorax, the thrust of the abdomen. Most species have distinguishing details.

The chrysalis of the Monarch, who forms the inverted question mark, looks like a jade earring. Near the light green top, an elegant band of gold is underscored with a thin black line. More highlights of gold decorate the bottom half. The pupa of the California Sister, similarly shot with gold, or the Dimorphic Bark Wing, with a silver oval on its dark green thorax, can also be described as jewelry. No one fully understands the purpose of this glitter. Perhaps the pupae gleam to warn off predators. Perhaps their reflectance camouflages them in the light and dark of a sunlit branch. They may be trying to look like metallic beetles. They may be imitating raindrops.

Other boldly colored pupae are the bluish white Baltimore Checkerspots with their orange bumps, black dots, and dashes. The chrysalides of some butterflies in Australia are a startling bright orange. A species in Costa Rica

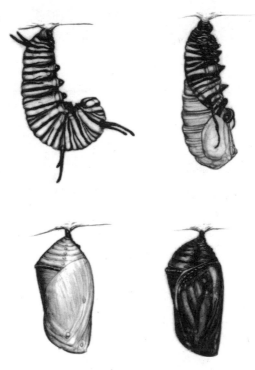

Monarch metamorphosis

is described as chrome-colored, like small car mirrors, with the wingpads edged in red.

But these are exceptions, for the fortunate and the bad-tasting. To a bird or lizard, most pupae are a convenience food: nicely packaged, immobile, full of nutrients. They are the original energy bar.

So the chrysalides of the Zebra butterfly look like brown,

drying leaves; and those of the Painted Lady, flakes of rock. The pupa of the European Orange-tip can be mistaken for a thorn growing from a twig.

In some species, in an individual prepupal caterpillar, the background color determines the color of the chrysalis: green for a green surface, brown for brown.

The pupa is carefully choosing its dress.

The pupa is playing a waiting game of hurry up. Sometimes it jerks its abdomen at the threat of a predator. Sometimes, in some species, it makes a clicking sound by using a file of teeth against an armored plate. Some pupae can hiss or squeak or give a vibrational pulse to warn away attackers or to signal ants. The pupae of skippers in the roots of yucca trees move up and down, awkwardly, in their long burrows.

But mostly the pupa is silent, still, intent.

What is it doing, exactly?

Many of the changes started taking place before pupation. The wings of a butterfly begin as early as the first larval stage, or instar, as thickening cells in the thoracic segments. These cells become two pouches called wing buds, or imaginal disks. By the last stage, the fifth instar, each pouch has folded in upon itself to make a four-layered structure corresponding to the future upper and lower surfaces of the adult wing. A pattern of veins is established. A blueprint of the wing is forming, down to the smallest eyespot.

Other adult structures have also started to grow underneath the larval skin. After the caterpillar finds its resting spot, as it hangs in its prepupal stage, these new adult parts—the antennae or the proboscis for sucking up nectar—move to the surface. The caterpillar's color may change. Swallowtails turn brown.

By the time, as Nabokov sighed, "all the skin has come off, thank God," and the hard bared surface is revealed, the work of metamorphosis is largely done.

In the first half of pupation, the wing disks grow until they are the size of adult wings confined into the small space of the chrysalis, their surfaces compressed, like a rubber balloon not yet inflated. The scales on the wings develop. Pigments are synthesized to fill in the waiting design. It's a paint-by-numbers set. Just before emergence, the final touches are added.

In a Buckeye, the rings around the eyespots are colored yellow.

From the beginning, cells in the caterpillar have been preparing the way, genes flicking on and off. Here, at last in the hardened pupa, they resemble a thousand pinball machines. Bang, clang, rebound! This is pinball wizardry, chaos controlled, nothing random. The simple larval eyes dissolve. The butterfly's complex compound eyes grow from other cells. Legs lengthen and add segments. New muscles develop, some for flight. The huge, dominant stomach shrinks. The sexual organs appear. Eggs may mature in the female, sperm in the male.

Whistle, flash, ring! Everything is rushing, propelled to the right place at the right time to do the right thing. Cells die and are reabsorbed, cells divide, cells restructure. You're a winner!

The time spent in the chrysalis, from liquefied caterpillar to adult butterfly, varies in different species from days to weeks.

In climates that get very cold or very hot, the pupa may delay its transformation and hibernate during the winter or summer months. Some pupae can wait for the right signal, whether warmth or rain, for five to seven years.

A bag of goo crawls on a leaf, obsessed with eating. It hangs upside down. It becomes something else. A butterfly is born, a bit of blue heaven, a jazzy design.

It is a gesture of beauty almost too casual.

We are storytelling animals. The peak is high and white with snow. Who has not seen God in the mountains?

Do we create the story or resonate with it?

The story of the butterfly has been read the same by people around the world.

When the Hindu god Brahma watched the caterpillars in his garden change into pupae and then into butterflies, he conceived the idea of reincarnation: perfection through rebirth. The Greeks used the word *psyche* for butterfly and for soul. Ancient images on Egyptian tombs and sarcophagi show butterflies surrounding the dead. In the fifth century, Pope Gelasius I made a pontifical declaration comparing the life of Christ to that of the caterpillar: *Vermis quia resurrexit!* The worm has risen again. In Ireland, in 1680, a law forbade the killing of white butterflies because they were the souls of children. In Java, in 1883, a migration of butterflies was interpreted as the journey of the 30,000 people killed by the eruption at Krakatau. In China, in the 1990s, single white butterflies were found in the cells of executed convicts recently converted to Buddhism.

The butterfly is a human soul. What could be more obvious?

After World War II, Elizabeth Kubler-Ross visited the barracks of a Polish concentration camp and saw hundreds of butterflies carved into the walls by Jewish inmates. "Once dead, they would be out of this hellish place," she wrote. "Not tortured anymore. Not separated from their families. Not sent to gas chambers. None of this gruesome life mattered anymore. Soon they would leave their bodies the way a butterfly leaves its cocoon."

Miriam Rothschild saw the same image drawn with orange chalk in Jerusalem: "It was a butterfly every Jew and Jewess knows by heart, the butterfly drawn by children in German death camps before they went to the gas chambers. Fifteen thousand children had been incarcerated in one particular camp; only a hundred survived. It is the emblem of escape from the greatest sorrow the world has ever known."

Butterflies, death, resurrection. Besides the Victorian collectors, few cultures have been as obsessed with butterflies as the nobility of ancient Mexico, the people who turned ritual sacrifice into a state-sponsored art form. On the first day of each month, hundreds of Aztec subjects—children, captives, and slaves—were killed. On special occasions, thousands died.

In the sixteenth century, high-ranking Aztecs carried bouquets of flowers. Everyone knew that it was impolite to smell these flowers from the top, since that was reserved for butterflies, the returned souls of warriors and sacrificial victims. Toltec and Aztec shields were often decorated with butterflies, possibly a reference to the goddess of love, Xochiquetzal, who held a butterfly between her lips as she made love to young men on the battlefield. Her kisses were their assurance that they would be reborn if they died that day.

Xochiquetzal was the mother of Quetzalcoatl, the god of life, who suggested that the Aztecs offer tortillas, incense, flowers, and butterflies instead of human hearts ripped from living chests. The idea never caught on. In a single celebration, the Aztec ruler of Tenochtitlán once sacrificed 10,000 prisoners, marching them up temples awash in blood. Theoretically, the victims may have all become skippers, swallowtails, Monarchs, and Madrones.

From what we know about a caterpillar's life, the idea of war and sacrifice juxtaposed with butterflies is not unreasonable. Entomologists deal in the archetypical symbol of spiritual transformation. They are rarely sentimentalists.

A common experience for anyone collecting butterflies is to keep watch on a chrysalis, in happy expectation, only to see a parasitoid emerge, usually a wasp or group of wasps.

As Philip DeVries pointed out, "Incidentally, it is during the noticeable wandering phase that many casual entomologists take note of caterpillars and subsequently confine them in containers." At this point, he added, it is just as likely that something else has found the caterpillar first.

It happened to me once. I was eight years old. My third grade classroom stood next to a large group of trees, possibly mulberries, which erupted in the spring with teeming nests of caterpillars, possibly silkworms. Rumor had it that our teacher once punished a child for keeping a caterpillar by cutting it in half with her yellow ruler. I rushed my caterpillar home and put it in a shoebox with lots of mulberry leaves. You can imagine my delight when it spun a cocoon.

Later, when the cocoon broke open, you can imagine my dismay.

All this underscores the miracle of what can and does happen regularly. The caterpillar forms a question. The question is answered when a Queen, cousin of the Monarch, emerges from her chrysalis just before dawn with a swollen-looking body and pathetic, wet, crumpled wings.

She needs gravity to help her, and so she climbs to where her wings can hang down as she pumps blood through their veins to expand and harden them. She needs to dispose of waste products, too, and she emits a brownish red fluid. She needs to remove dead cells and skin from her antennae. She needs to zip together the two halves of her proboscis so that it will function as a perfect drinking straw.

Tentatively, she moves her head and thorax, which are beautifully patterned, white spots on black. Small bumps, or palpi, near her clubbed antennae may be used later for cleaning and brushing. Her thorax and abdomen are her toughest parts, capable of being grabbed, tested, and rejected by a predator. Like all insects, she has six legs; on her, the front pair are reduced.

Perceptibly, second by second, she gains strength. Her large wings are a russet orange with a black scalloped border. The top pair are called forewings; the lower pair are hindwings. She moves in place, lifting her legs, weighing perhaps a third of what she weighed as a caterpillar.

Once she was a stomach attached to a mouth. Now she is designed to float through the air.

Her old obsession was food. Her new obsession is mating and laying eggs.

In an hour or so, she is ready to fly away.

Vermis quia resurrexit!

FIVE

BUTTERFLY BRAINS

*F*LOWERS AND BUTTERFLIES HAVE A BUSINESS relationship. Butterflies need the energy in the nectar that a flower produces to attract pollinators. When a butterfly uncurls its proboscis to probe for that nectar, grains of pollen, or male sex cells, from the flower's stamen are caught on the insect's body. This packet of sperm is carried to another flower, where the pollen has the chance of dislodging onto the female stigma and fertilizing that flower's eggs.

When flowers trade with butterflies, they want a smart business partner who is not too smart (not like the bumblebee, notorious for entering from behind the base of a flower and stealing nectar without picking up pollen) and not too dumb (not like some ants who take nectar and carry off pollen, which they inadvertently, chemically, render sterile). When flowers talk, they want a good listener, someone who can learn the rules and stick to them. They want someone who will pick up a load of pollen and fly off to another flower with a similar color, a similar shape, and a compatible female stigma.

Flowers and butterflies have worked out a deal that, to be honest, is not as sophisticated as the deal worked out by flowers and some other insects. Among biologists, there is a hierarchy of smart bugs. Honeybees are at the top.

"Honeybees are seen as the intellectuals of the insect world," says Martha Weiss, a biologist who studies butterflies. "They are under strong pressure to collect nectar and pollen efficiently and get their hive through the winter. They have that aura of the 'busy bee.' They have fabulous comb architecture. And they have this menacing sting, which I think adds to their glamour."

In one experiment, honeybees were trained to select every morning, between 9:30 and 11:00, the lower right petal of a blue flower and then, between 11:00 and 12:30, the lower left petal of a yellow flower. They were accurate about 80 percent of the time, equaling perhaps how the researchers might have done if given the same task. In real life, like some butterflies, honeybees "trapline" a series of flowers, which open and provide nectar only at certain hours of the day. Their ability to learn is embedded in time.

In another experiment, for three collecting flights, honeybees were given different scents at nine times only

twenty minutes apart. Each scent brought a different reward. The next day, the bees could pick out the right scent
at the right moment, nine times out of nine.

Honeybees, like ants and many wasps, are social insects
living in highly organized nests. Scientists have assumed
that learning ability would be greatest in these species,
who had to evolve ways of communication and interaction.

Entomologists also know more about bees because bees
are easier to study, easier to rear, and easier to keep as a
sustaining colony.

"But a lot of other insects are just as smart," Martha
says. "They're underappreciated. They're stereotyped. Butterflies, for example, are beautiful. They bask in the sun.
They go around sipping nectar from flowers, mating, and
laying eggs. They look lazy and ornamental. In fact, they
are, smartly, doing just what they need to do."

For solitary insects like butterflies, a sophisticated learning style might be even more important, since individuals
must shoulder alone all the burdens of life: foraging, mating, finding shelter, egg laying. Most of these tasks must be
done quickly, in days or weeks, in environments that are
often unpredictable.

Martha was first interested in flowers, not butterflies.
She wondered why so many species of flowers change
color as they grow older. The white spot on the blue petal

of a lupine turns purple. A field of white lilies becomes pink and red. Another white flower loses its ring of yellow.

Lantana camara flowers are part of a bushy inflorescence, a group of small flowers on one plant. On the first day of opening, the lantana flower is lemon yellow. On the second day, the flower turns orange. On the third day, the flower is red. Martha now knows why these and other flowering species, in over two hundred genera, change color with age. They are signaling their pollinators that either they are empty of nectar or likely to be empty given the passage of time. They are also, hopefully, already fertilized. The pollinator would do well to probe a younger, unfertilized flower on the same *Lantana camara* bush.

The lantana bush wants to have as many of its flowers probed and fertilized as possible. The butterfly wants to gather as much nectar as possible while doing the least amount of work.

A flower could show that it is already fertilized by shriveling up and dying. But if reproductive changes are still taking place, parts of the flower may still be useful.

Also, as Martha says,

Many pollinators are visually oriented. Something catches their eye. They make a decision to come over. A showy display of flowers is a long-distance flag, attracting butterflies from far away. Once the pollinator is up close, the plant needs to send another signal showing which flowers to probe. The lantana bush keeps its flowers for three days. If the plant didn't signal which flowers were rewarding and which were not, then the pollinator might be wasting her time, and she would get mad and leave.

Martha asked herself why the flowers were changing color and answered that question. Next, she wondered about the butterflies who, for the most part, were sticking their proboscides into the nectar-rich yellow lantana flowers, not into the orange or red. Was their response instinctive or learned or both?

Was all that trash talk about butterflies true?

Experiments with insects can involve an arts-and-crafts approach. It's not hard to see Martha, the mother of two children, happily making daisylike flowers out of fringed strips of paper glued to a plastic pipette tip. Over two inches in diameter, in colors of yellow, blue, red, green, orange, and purple, the artificial flowers were placed randomly six inches apart on a brown cardboard background.

By this time, Martha had collected the larvae of Pipevine Swallowtails and had raised a generation of these butterflies, their dark blue–tailed wings spotted red and yellow. She let forty-six of them, one by one, into the cage of fake flowers. The newly emerged, or naïve, Pipevine Swallowtails had never seen flowers before. Martha and her colleagues watched what they did. Most, very clearly, preferred yellow. Then blue. Then purple.

Next, Martha went to bushes of *Lantana camara*. Isolated from feeding butterflies, all the flowers of this bush will keep their nectar even though they still change color as they age, from yellow to orange to red. Martha removed all intermediate or orange flowers and manipulated the bushes into three types, using tiny paper wicks to remove nectar. One bush retained its natural pattern in the wild of nectar in yellow and no nectar in red. One had the reverse, no nectar in yellow and nectar in red. One had no nectar in either form.

Now Martha had her fun, watching how the inexperienced swallowtails treated each bush, and how they reacted when they were moved from one kind of bush to the next. Although the butterflies instinctively chose yellow, they would switch to red after learning that the red flowers were the ones with nectar. This took about ten flower visits. When the pattern was switched again, the butterflies switched again. They adjusted. They readjusted. This further confirmed what researchers had been saying: Butterflies had the brains to be flexible.

In a later experiment, Martha and her colleague Dan

Pipevine Swallowtail nectaring

Papaj trained female Pipevine Swallowtails to associate the color of green, blue, yellow, or red with an extract from the leaves of pipevine, the plant on which the butterflies oviposit, or lay their eggs. The same individuals were then trained to associate a different color with a reward of nectar. Offered an array of paper flowers, most butterflies went to their assigned colors, uncoiled their proboscides at the color for nectar, and drummed with their forelegs or curled their abdomens (pre-signs of oviposition) at the color for egg laying.

From this, we know that Pipevine Swallowtails can retain and respond to the meaning of two different colors in two different contexts.

In a further experiment, Dan and a student trained females to lay eggs on red and blue and to keep their innate preference for green; that's three different colors the swallowtails were able to understand as a signal for egg laying. This achievement was impressive, since the swallowtail's memory could not be associated with the activity of feeding or the activity of egg laying but had to be organized in the context of a single activity and two other memories.

In all his work with butterflies, Dan has found that they have, well, individual personalities. "I do not mean to imply anything about consciousness," he hastens to say, "or to draw unwarranted connections to human existence."

But butterflies are not robots, each acting like the other. In experiments in which Dan tagged and followed butterflies in the field, he found he could identify some insects "on the wing" because of their behavior, not their tags. Individual butterflies searched for food or host plants in a certain way or in certain places or with a certain intensity.

"Initially there may be genetic differences and later differences in nutrition," Dan says. "But in addition, butterflies are almost certainly learning a great deal about their host environment, about occurrence and types of predators, about how to fly best given their size and weight, about habitat structure, and on and on. Each butterfly is having slightly different experiences on each count, so that any given butterfly's experience is unique."

If we all quit our jobs and went out right now to observe butterflies, we would discover all kinds of things. Obviously, it is not just Pipevine Swallowtails who juggle memory. Other species must be equally clever.

Cabbage Whites also pick out the right color for the right reward, after just one experience. There is no reason to doubt a similar intelligence in the Peacock or the Cloudless Sulphur or the Painted Lady. Or, for that matter, in a sheep blowfly, who can associate rewards and colors about as fast as a honeybee.

In various experiments, Cabbage Whites also learned how to find nectar more efficiently in the complicated shape of a bellflower or trefoil. The first visit took up to ten seconds of fumbling. By the fourth visit, the butterfly had cut his time to less than half.

When a Cabbage White had to switch flower species, however, the insect was briefly nonplussed, even when the second species was one the Cabbage White knew well. This "interference effect" may promote the kind of loyalty or constancy the flower needs. Flowers don't want to have a shape too easy to learn and relearn. The trefoil wants the Cabbage White to have a reason to keep probing more trefoil, the species the butterfly has already mastered, and not to switch to some silly-looking bellflower.

Butterflies switch anyway and lose efficiency in the process. Skippers moving from one flower species to another increased their handling time by less than a second. Is that important? When you are skipping about to hundreds of flowers a day, do hundreds of less-than-a-seconds add up to a lot? Or are they too little to matter?

It's something to think about, even if your brain *is* smaller than a pea but bigger than a poppy seed.

In a competitive world, we have to wonder: Who's the smartest butterfly around? Most people point to the *Heliconius* genus, a well-studied tropical group whose caterpillars feed mainly on passion vines. Some of these butterflies are called longwings, or postmans, for the regularity in which they visit a trapline of flowers. Most are brightly colored, with dashes of orange, red, yellow, or blue, obvious signals of unpalatability.

Longwings can live to be old for a butterfly, up to eight months. Because they spend less of their life cycle as caterpillars than most groups, they eat less and gain less weight. When the adults emerge from their chrysalides, they still must find 80 percent of the nutrients they need to make and lay eggs. Perhaps for this reason, their foraging behavior is complex. They learn a daily route of plants, and they are one of the few butterflies who eat pollen. This involves processing the grains in the proboscis: mixing and soaking them in saliva, agitating them by coiling and uncoiling the tongue, and then drinking the liquid now rich with amino acids. Tested against other species, pollen-eating butterflies manipulated lantana flowers faster and probed them more thoroughly.

Butterflies in this genus have good memories. They remember favorite flowers. They remember favorite roosting sites. They remember to hold a grudge, avoiding spots where some scientist captured them days earlier.

In one laboratory, only the *Heliconius* could remember not to fly into the fluorescent bulbs.

They score about a 2 on the SAT tests.

Like most scientists, Martha Weiss follows questions, one leading to the next. She has moved from why flowers change their colors to how butterflies learn color to the regularity of leaf shelters made by skipper larvae to the mechanisms with which skipper larvae eject their frass to how other animals get rid of their waste. She is now working in a field she calls "defecation ecology."

"Caterpillars are a good group to study," she says, "because their frass is relatively innocuous. But I've also gotten interested in looking at how other animals dispose of waste. Birds, for example."

This has nothing to do with butterfly brains. It has everything to do with the human brain.

SIX

BUTTERFLY MATISSE

More than any other group of animals, butterflies look as if they were designed in art school.

The Zebra Longwing with stripes of black and white like the fabric on my father-in-law's couch.

The underside of the Two-tailed Pasha, swirls of chocolate brown, white, green, a row of orange dots, metallic blue, scallops of yellow. A flying scrap of paisley. Incense and beads. Joan Baez.

The abstract art of a tropical Postman. The deco art of a Black-veined White. The checkered-tie quality of fritillaries and crescents. The morpho as a stained-glass window. The Southern Festoon as a circus poster.

Chevrons, zigzags, teardrops.

The colors and patterns just don't seem natural.

Butterflies are two pairs of wings flapping about in broad

daylight. They don't have teeth or claws. They can't fly very fast. Their abdomens make for a quick snack.

Art is their defense.

Like old-fashioned sandwich boards, butterflies can be seen from two sides. Open, spread out, their forewings and hindwings have an upper design. This is what you see when the butterfly basks in the sun or glides through the air. Closed, clapped together, the forewings and hindwings have an under design. This is what you see when the butterfly rests on a leaf or stops to drink nectar. Almost always, the two patterns, upper and under, are very different.

In butterflies, as with caterpillars and pupae, bright colors and a simple pattern are a warning that is easy to remember. An inexperienced blue jay who eats a Monarch will next be seen retching, vomiting, jerking its head, fluffing its feathers, wiping its bill, and closing its eyes in the expression of a blue jay calling out to its deity.

But art is more than a skull and crossbones.

Art can be a magic cloak.

The ripples and swirls on a Gray Cracker blend into the ripples and swirls of tree bark. The Marbled Fritillary, spotted in patches of light and dark, disappears in a woodland of spotted sunlight. The leaf-mimic holds a pose and becomes invisible.

Some leaf-mimics even have areas that look like patches of white, simulating light through a leaf's tear.

Art is deception.

Art is also distraction.

The bands of color on a butterfly's wing lead a predator's eye away from the head to the more expendable tail area. Typically, the stripes and curves end in the eyespot of a false head, a representation sometimes abstract, sometimes realistic. The bird jabs at the "eye," bites off a bit of wing, and the butterfly escapes. Even a large wing tear usually doesn't stop the insect from flying away, from a few more hours of life, another chance to mate.

Long-tailed Blues add the appearance of antennae. Hairstreaks at rest have the most convincing "dummy head" with eyespots and two streamers that move antennae-like in the breeze.

Some eyespots are deliberately large and contain pupils, quite different from the butterfly's compound eye. These are vertebrate eyes. Cat eyes. Raptor eyes.

The Peacock butterfly has two purple, black, cream, and red eyespots on its upper forewings and two purple, black, and cream eyespots with slitted pupils on its upper hindwings. Faced with a predator, the Peacock opens this display and makes a hissing sound by rubbing the veins of its hind- and forewings together.

Variegated Fritillary

Snake eyes. Devil eyes.

The pattern of the butterfly's underside can also camouflage, distract, or startle. The opened wings of a Painted Lady are pure advertisement: black, orange, *think again*. But once she is perched, her wings closed, the browns of her underside dissolve easily into backgrounds of branch or earth. If discovered, the Painted Lady moves her forewing forward to show a previously hidden patch of orange. The color flashes. Then the hindwing hides it again. The blue jay is startled and his search image becomes confused. *Am I looking for something colored or for something camouflaged?*

For a long time, biologists thought the colors of a butterfly had evolved so that sexes could recognize each other. In many species, males and females look different. The male of the Northern Blue is blue; the female is brown. Charles Darwin was convinced that lovelier blues were determined by female selection, the females preferring brighter colors and males brightening up to oblige.

But many females do not seem to distinguish much between the visible colors. In experiments with sulphur butterflies, researchers dyed the wings of males green, red, blue, or orange. Females still recognized the males, probably through scent, and mated with them. Only the reflectance of ultraviolet light (a color the human eye cannot see) seems in some species to influence a female's response.

Males, on the other hand, do use color to find an appropriate mate. Orange Sulphur males will swoop down to investigate paper dummies, especially those colored yellow-green to resemble the underside of female sulphurs. The males of other species may follow the flip of a red hair ribbon or the glimpse of a blue dress.

Moreover, males use color to avoid each other. For a male Orange Sulphur, the wing of another male reflecting ultraviolet light is repellent. The sight of such an ugly wing inhibits the approach of Orange Sulphur males, as well as that of males from a related species such as the Common Sulphur. These males don't want to get to know each other. They don't want to waste time swooping down to court another male.

Art is communication. Art is the flare gun of possible sex. Art is a nasty memo between guys.

The word *lepidoptera* comes from the Greek *lepis,* scale, and *pteron,* wing. Butterfly wings are made of two flat, thin sheets pressed together and supported by a system of hollow tubes or veins. The wings, upper- and underside, are covered with overlapping rows of scales. A typical scale has a bladelike body, a smooth bottom, and a ridged top; a stalk fits this body into a socket on the wing surface. Scales are found all over the butterfly, including the head, abdomen, legs, and thorax. Scales are modified hairs; their original use, mundanely, may have been for insulation.

Each wing scale is one color. The larger pattern is a mosaic of these. The color can come from pigments in the scale, from its structural architecture, from the effect of overlapping scales, or from a combination of all three.

Pigments in a scale reflect red, orange, and yellow, some blues, greens, and purples. Melanins are the most common pigment, their molecules absorbing most of the light spectrum to show only black or shades of brown. Dark colors "keep" more light and help the butterfly warm up quickly. At high altitudes, butterflies are often dark or have dark markings.

Art is heat.

The physical structure of a scale causes light to scatter or diffract. The matte white of a Cabbage White butterfly is created when light is scattered in all directions by the corrugations on the scale's top surface. Changing a scale's structure, raising or angling its ridges, filling its interior with a crystalline latticework or with thin, evenly spaced layers, results in varying colors and effects: the iridescent blue of a tropical morpho, the satin sheen of a Pipevine Swallowtail, the emerald of a Green Hairstreak.

The butterfly's palette lies in a micron's shift, in the thickening of a ridge.

The effects harmonize. The architecture for iridescent blue combines with red pigment for metallic violet. The structural white of some butterflies needs uric acid, a waste product, to produce a certain shade of cream. Pearly white depends on the overlapping of scales.

We love butterflies, in part because we can know them so easily. Most of the 18,000 species have unique wing patterns that distinguish them from all other species. Get a guidebook, take a few days, and you'll be able to recognize on sight the major groups in your area. That's a Painted

Lady. That's a Dotted Checkerspot. That's a Southern Dog-face. Butterflies make us feel smart.

Butterflies make us feel smart because butterflies themselves are so well organized. Researcher Frederik Nijhout reminds us that each element on a butterfly's wing is "not just the random dot or stripe seen in the coat colors of mammals." Rather, they are part of what scientists call the nymphalid ground plan. The basic units of these spots, bands, and borders change in color, shape, and number, but their place on the wing is fairly constant.

The patterns in a butterfly's wing, Nijhout says, are "as consistent and fundamental" as the skull bones and limbs of a bobcat or an elephant or a whale. They show the "diversity that can be achieved by permutation and recombination of a relatively small number of basic units."

Mix, match, add, subtract.

Art is a numbers game.

Get a guidebook, take a few years, and you'll still make mistakes. Butterfly identification has an initial, deceptive simplicity. You feel confident. You swagger. You like the sound of your own voice.

Then you look again.

Individuals in a species vary naturally, an eyespot slightly larger, a color brighter.

The males and females of a species can be strikingly dissimilar.

So can genetic morphs or forms within a gender. The female Tiger Swallowtail may emerge from her chrysalis striped, like the male, or darkly colored to mimic the dark blue of the bad-tasting Pipevine Swallowtail. Yellow Tiger Swallowtail females produce yellow daughters. Black Tiger Swallowtail females produce black daughters.

Within a species, individuals also show mutation or change. The pressure of natural selection is the way something in the environment or some other condition causes some individuals to die and some to survive and successfully reproduce. The result of natural selection can be that an entire population or group now has the adaptive strategy (mutation or change) of the successful individual.

In a single species, butterflies who live in a range of habitat can vary in appearance and produce different populations or geographical races, each race better adapted to its environment. In the United Kingdom, the Large Heath becomes paler, with fewer eyespots, the farther north it is found. Populations or races at either end of a range can be confused for two different species.

Species can also produce generations of distinct morphs in the same place at different times of the year. European Map butterflies that hatch in the spring look like fritillaries; those that hatch later in the summer resemble White Admirals.

The dry, cooler-season form of an African butterfly has tiny eyespots on the underside. The wet, warmer-season form has large eyespots. In the dry, cool season, when the butterflies are mostly inactive, big eyespots may make them more susceptible to attack by birds and lizards; a less noticeable pattern is a better defense. In the wet warm season, a distracting predator-deflecting eyespot might be preferable.

In these species, different wing patterns depend on the temperature, either when the larva is developing or when the adult is forming inside the chrysalis. For the African butterfly, smaller eyespots occur when the chrysalis is cooler and larger eyespots when the chrysalis is warmer.

In response to changing seasons, these butterflies produce individuals whose offspring can respond to that change; those who cannot die out. When scientists raised these butterflies in a laboratory under a constant temperature, however, and artificially bred for two lines of small and large eyespots, the insects lost their ability to develop the two forms after fewer than twenty generations. They became less flexible. No matter how cold or how hot it was, one line produced small eyespots only and one line produced large eyespots only.

One Tiger Swallowtail is black. On the same flower, another Tiger Swallowtail is yellow. A Speckled Wood butterfly is lighter in Scandinavia than in North Africa. A species produces red butterflies in the wet season and blue in the dry. A species darkens in response to air pollution. A species gains an eyespot. A species loses one.

Art is a constant dialogue with the world.

SEVEN

LOVE STORIES

SOMEWHERE IN EUROPE, LET'S SAY FRANCE, a male Grayling butterfly sees a shape that he likes: a squarish rectangle making a skipping motion through the air. For this male, the color of the shape is not very important, although the Grayling will respond with a bit more excitement to red or black. Sometimes the shape is a female Grayling, a brownish gray butterfly with a pattern that allows her to hide well against bare ground. Sometimes it is a leaf or a big bee or a scrap of paper. The Grayling pursues hotly, filled with hope.

If the object of the pursuit is not a piece of candy wrapper, and if the female is receptive, she will land. The male follows and turns to face her. He shivers. He twirls his antennae. He expands his wings and fans them. He lays his antennae on the ground. He arches his body in what has been described as an elegant bow and hoods the female's antennae with his forewings. He sprinkles the receptors on her antennae with a chemical pheromone that even scientists call love dust.

Now the male tries to touch the female with his abdomen. If the love dust has worked its spell, she will raise her wings and relax. (If she does not want to mate, she will tense and flutter her wings rapidly.) Two valves, or claspers, on the male open to expose his penis and to squeeze the female's abdomen, revealing and making accessible her genitalia.

It takes a few seconds to fit his parts into hers. Locked into place, the pair stay bonded for about an hour.

The Tiger Swallowtail patrols a woodland lane or a moist valley bottom or a riverbank. He is on the prowl. He swings by the best flowers in town, but he is most interested in those trees and bushes he remembers from childhood, the plants he ate as a caterpillar. Females just emerging from their chrysalides and females looking for a place to lay their eggs are likely to be nearby.

Typically, butterfly males either patrol for a mate or perch and wait for one. When the host plant for the caterpillar is easy to find, the male tends to travel; females are easy to find, too. When the host plant is scattered or rare, males tend to establish and protect a territory, often a conspicuous rendezvous point, and wait for the female to find

Mating Graylings

them. Males may also defend an actual patch of host plant or nectar source. Species adapt to local conditions, and individuals adapt, too.

"Hill-topping" butterflies such as the Black Swallowtail think the highest spot is the best one. Aggressively, they chase away males of the same species. Territorial butterflies often engage in spiral flights as one male goes under the second male and then up (the same strategy used to get in front of a female) and the second male repeats the pattern, and under and up they both go, under and up, under and up. A few species buffet each other vigorously.

Others battle with sound: The Gray Cracker makes a loud, impressive noise with its wings. The original holder of the territory does not always have the advantage but will give way to greater size or experience or determination.

Some males perch and set up a territory for as long as they live. Others move on, an afternoon here, an afternoon there, a series of kingdoms and battles.

The Tiger Swallowtail patrols; he's looking for that seductive shimmer of yellow. Perhaps the black stripes titillate him, too, or the row of blue dots on the lower wing, or that red splash near the tail. It is true, the male has to remind himself, that some female Tiger Swallowtails are darkly colored to mimic the bad-tasting Pipevine Swallowtail. Occasionally, the male does not recognize these females and passes them by. He regrets that.

Like all male Tigers, he emerged from his chrysalis earlier than the females, days earlier. This has given him time to establish the best patrol, to eat, and to puddle. To puddle is a verb for the Tiger Swallowtail, an activity in which he and other males gather at a wet site, preferably an area mixed with urine or dung. Like drinking buddies at a local bar, they chug minerals and salts and, sometimes, amino acids. If the soil has nutrients but is dry, the Tiger will regurgitate fluid to moisten the ground and then suck the moisture back up his proboscis. If he's really lucky, he'll get to puddle at moist feces. Or maybe something dead.

Mostly, however, the male has been waiting.

For her. For love. For destiny.

When he finds her, he will flutter, and she will flutter, and sweet pheromones will scent the air. Even a human passing at the right moment might pause and sniff, and sniff again. Honeysuckle? Lavender? Jasmine? The pheromones of butterflies have long co-evolved with the sexy scent of flowers promising food and drink (the flowers desiring sex, too) and we have long since taken these scents for ourselves, for our perfumes and our colognes, for our own longing.

The butterflies land and join. In the business of copulation, the male swallowtail may have an advantage: simple eyes on his genitalia that respond to light. When he is correctly aligned with the female, the cells that sense light are blocked, and he knows he can start the next step.

The male transfers his sperm in a thick-walled sac mostly made of proteins he acquired as a caterpillar, as well as other nutrients from nectar and puddling. This packet is the spermatophore, or sperm carrier, and it can be from 4 to 8 percent of the male's body weight. The female will use this nuptial gift to support herself as she searches for a place to lay her eggs. In some species, the proteins in the spermatophore are needed to produce more eggs.

The bigger the gift, the better. The bigger the gift, the longer the female will wait until she mates again, and the more chance the male has that all his sperm will be used before being replaced by another male's. In butterflies, the most recent sperm has precedence; last guys finish first.

A spermatophore is the male Tiger's main parental investment. Some butterflies do more for their offspring. The Cabbage White male passes along a chemical that the female spreads over her eggs as they move through the oviduct. The pheromone signals other female butterflies not to lay eggs nearby. This food is meant for this male's larvae.

The male Tiger disengages. His second gift is a small plug that he leaves inside the female. He is trying to prevent her from mating again. Used in many species, mating plugs vary in size and effectiveness. Eventually, this one will get lost or broken. There is some thought that the swallowtail female can remove it herself.

The Queen butterfly resembles both the Monarch and the Viceroy: All three species are warningly colored in orange and black. This mutual mimicry enforces the idea that all three species can taste bad. A bird biting into an unpalatable Monarch remembers the experience when he sees an unpalatable Viceroy or Queen. The resemblance helps the Queen survive but makes it visually harder for males to find compatible females. In species who use mimicry as protection, chemical signals in mating become even more important, and more complex. Along with appropriate sex and species, these signals also communicate age, fitness, and mating history.

Personal ad #24: M seeks F, 23–35, no pets, no parasites, no kinky hobbies, must like kids and nectar.

The male Queen begins by going shopping. For his perfume, he needs the ingredients found in the dried, withered leaves of alkaloid-containing plants. The Queen regurgitates on the dead leaf and drinks the fluid that now contains the alkaloids released when the plant tissue breaks down. To start this process, butterflies have been known to scratch at undamaged plants. In the laboratory, males denied these plants have trouble mating. In the wild, swarms of Queens and Monarchs can be found jostling and pushing each other for a good place on a broken branch of wilted, dead leaves.

In some butterfly species, alkaloids found in plants are absorbed and used as a way for the adults to become poisonous or bad-tasting to their predators.

Queens and Monarchs, however, depend mainly on another toxin, the heart poisons, or cardenolides, found in milkweeds. For the Queen, males use the toxic alkaloids to help synthesize the pheromone that attracts and lures a willing female. The male stores that pheromone in a handy gland on his hindwing.

Male Queens also have brushlike organs, or hairpencils, tucked away in their abdomens. When the male sees a female, he inserts his hairpencils into his hindwing gland and gathers his personal scent. Now he flies under and ahead of the female, expands his hairpencils, and dusts her. The more alkaloids the male has ingested, the better he seems to smell, and the more clearly he seems to signal: I am fit, I am able, I have a big nuptial gift.

The male's chemical bouquet contains an inhibitor against flight and a glue to keep the dust on the female's antennae. Females courted in the air land. The hairpenciling continues. A female says yes by closing her wings and giving the male access to her abdomen. He cuddles closer and palpates her antennae as they join.

She says no by fluttering her wings. Now the male tries dropping on her repeatedly and forcing her back into the air, where he will repeat the entire process. These second attempts may or may not be successful.

Once their genitalia are locked, the male rises and carries the female in a postnuptial flight. They may be together for as long as eight hours, and he prefers a more private place. (In some species, the female is larger and carries the male. Sometimes, after mating is finished, a female will fly off, dangling her partner, simply in the hope of dislodging him.)

During copulation, the male passes on his spermatophore—the sperm and nuptial gift—which includes the alkaloids collected earlier from plants. Possibly the female uses these chemicals to increase her toxicity or she may pass them on to her eggs, to help protect them.

The female Queen mates as many as fifteen times in her life. She'll produce and lay a number of egg batches fertilized by a number of males.

Let the male collect the alkaloids. The female has enough work to do.

For the Monarch, the story is less gallant. These males have very small hairpencils that lack the scent needed to seduce a female. Monarchs usually mate in large colonies that have overwintered together. There are plenty of females and not much risk of mating with the wrong species. Instead of courtship, the Monarch performs an aerial takedown, dropping on the female and forcing her to the ground. There he touches her with his antennae, she acquiesces, and he carries her off in flight.

Miriam Rothschild described the Monarch as a thug and a "prime example of nature's male chauvinistic pig."

Monarchs can be seen drinking quantities of dew on days when they are busy mating. Their spermatophore can be 90 percent water and can equal 10 percent of the male's weight; the bigger and wetter the spermatophore, the longer the female will resist re-mating.

Personal ad #189: M, forceful type, wants F any age, minor role-playing, must enjoy airplane rides.

Make love, not war. Sometimes, it's the same thing. For most butterflies, the female can prevent rape by raising her abdomen in a posture of rejection and thus control access to her genitalia. Still, rape is not uncommon in some species, and it's the norm in others.

Male Zebra Longwings in the *Heliconius* genus keep watch on the passion vines on which caterpillars become pupae. These males scent out female pupae, inspect them regularly, and position themselves for when the young female emerges.

Or rather, just *before* she emerges.

Now the butterflies compete as to who will be the first to puncture the pupal skin and insert his genitalia. Larger males grab a position on the pupa itself. Other males circle the air and attempt to land. In some *Heliconius,* the female mates only this one time. In others, she can mate again. Occasionally, she is injured during the rape and dies.

Pupal mating probably evolved in populations of low density, where patrolling males looked desperately for unmated females. Someone increased his chances by finding and waiting beside a pupa. Pupal mating became the next preemptive strike. Some *Heliconius* males also mate with the pupae of different but related species; this interspecies mating kills the female and may have the effect of reducing competition on host plants.

Apollo butterflies have also abandoned the niceties. Males grab females in flight or capture them on the ground, finding virgins by smell as they hide in the grass. These females have external genitalia that are easy to access by force. After mating, the male secretes and glues a structure called a sphragis over the female's abdomen, a much more rigid and complex device than the small internal mating plug. This chastity belt is meant to last a lifetime. It is an extraordinary burden to carry around—heavy, awkward, and in the way of laying eggs. Only about 1 percent of butterfly species resort to this extremity.

Male apollos still attempt to capture mated females as well as virgins. Young mated females try to fly away or, failing that, struggle with their attackers. An older female is more passive and waits motionless as the male attempts to remove her sphragis. Many females have scarring around their abdomens where the male's needle-sharp penis has slipped and cut them.

In almost all apollos, the sphragis is an expanded hollow

organ, too big and smooth to grasp; in some species, there is also an encircling girdle with two sharp projections curving out like horns. Most attempts to dislodge a sphragis are unsuccessful, although occasionally a male can lever up the structure. A deformed sphragis produced by an older male running out of resources, or a very new one that has not yet hardened, can also be removed. In one experiment, 5 percent of the female apollos observed were reinseminated.

The sphragis, or chastity belt, and the spermatophore, or nuptial gift, are both resources that come directly from the body of the male. The males of species that produce a sphragis do not produce much of a spermatophore; they no longer need to, since the females of these species can no longer reject the male by moving their abdomens or controlling their internal structures. From the male's point of view, courtship and the nuptial gift have become unnecessary.

Why would a female let this happen? Rather, why would evolution favor females with external genitalia who are easy to rape and who can be "locked up" against further mating?

By and large, females want to re-mate. They receive more nuptial gifts and resources for their eggs. They replace depleted or degraded sperm. They may find a better genetic donor. They will, at least, get a range of genetic donors for a greater variability of offspring who can, in turn, respond to a constantly changing environment. Mating is tiring and even self-destructive. But for the good of their eggs, for the success of their young, for the longevity of their genetic line, females re-mate.

By and large, males do not want females to re-mate, since new sperm takes precedence over their own. Males may try to prevent females from re-mating by evolving more effective mating plugs or by offering larger spermatophores, or both.

Butterfly species that use a sphragis seem to have derived from species in which the males once produced a large spermatophore and an effective plug. In further competition with each other, males developed ways to remove that plug.

Females may have responded with external genitalia that were more difficult to plug and easier for other males to unplug.

Males may have countered this development with the sphragis, a structure that became increasingly complex and durable.

Males rose to the challenge posed by other males by evolving a tool kit that could help them dislodge a poorly formed or incomplete sphragis. The male genitalia of these species include augers for levering, sharpened tips, and large claspers. The males of species that capture rather than court may also have one enlarged claw with which they grasp the female. Both sexes tend to have extraordinarily tough wing membranes, better suited to the wear and tear of the mating struggle.

In short, effective mating plugs prompted the externalization of female genitalia, which increased the chances of re-mating but which also caused a cascade of male responses that ended in forced copulation and the sphragis.

At least this is the plausible scenario conceived by Bert Orr at Griffith University in Queensland, Australia.

I'm glad someone can explain it.

"There was a time in my life," Bert confesses, "when butterflies supplied for me all that I might otherwise have sought in art, literature, religion, and romantic love. Butterflies have grace. They have purpose. They bring their own literature in the form of their names, grand-sounding heroic Greek couplets, *Agrias sardanapolus! Parnassius autocrator!*"

At the height of his obsession, Bert was between the ages of twelve and fourteen. Still, even today, he struggles to express the potency of the relationship, "the pure beauty," he explains. "It's annoying, particularly for a scientist, to have something which is so palpable, so real, defy verbal definition."

Many poets have felt the same way.

Bert first began studying sphragides in the Australian Big Greasy. The male, patterned in red, white, and black, is larger than the female and violent in his mating. The female has a washed-out, greasy brown color that may help mated females hide from males. The female also uses the rejection signal of lifting her abdomen to reveal her sphragis; the male assesses whether mating is worth another try.

In the battle of the sexes, the genital arms race, there are always surprises. In one African butterfly, the female counter to the development of a sphragis is a long tube that extends into her body. Now the material excreted by the male to create a sphragis is used up instead to fill this tube, an internal stalk that does not effectively block further mating.

In some species, females "help" the male by evolving hooks and snags that keep the small internal mating plug in place. This allows males to invest material in a large spermatophore.

"Sexual conflict is hard to predict," Bert says. "A female move may require many male moves to counter it, or vice versa. It's this potential imbalance which occasionally leads to runaway evolution with some extreme and bizarre results. Like the sphragis. Sometimes, we can see the initial evolution of a sphragis or mating plug. It may take only a little to tip the balance in favor of one sex or the other."

One result of a sphragis is that females are left without a nuptial gift. Fewer resources for the female can mean less breeding success. This may be why the other 99 percent of butterfly species have evolved other strategies.

The mating of two Sharp-veined Whites can last up to twenty hours. During this time, the male transfers a chemical anti-aphrodisiac to the female, which she uses later to get rid of unwanted suitors. Sharp-veined Whites can be aggressive, and if the female has just mated, she would prefer to spend the next twenty hours sipping nectar and laying fertilized eggs, not fertilizing new ones. In addition, she has a large nuptial gift to digest.

The repellent chemical produced by the male seems to be effective. Even virgins painted with it are scorned. Eventually, the anti-love-dust wears off, and the female Sharp-veined White is ready to be courted again.

For a few days, the male and female have a mutual interest.

In the world of butterflies, this counts as another love story.

EIGHT

THE SINGLE MOM

YOU ARE WEIGHTY. YOU ARE FILLED WITH EGGS. Your abdomen drags you toward the earth.

You have wonderful eyesight, and you are proud of that: your wonderful compound eye made of hundreds of smaller eyes, each with its own lens. You are able, simultaneously, to see up, down, forward, and backward. You can see all the colors—red, orange, yellow, green, blue, purple, plus ultraviolet, plus all the colors mixed with ultraviolet, purple and ultraviolet, blue and ultraviolet, green and ultraviolet. You don't easily adjust for distance or see detailed patterns and shapes. But you know and recognize two very important shapes: You know the broad, oval leaves of the Texas Dutchman's pipe and the narrow, grasslike leaves of the Virginia snakeroot.

In the pine woodlands of east Texas, where you live, you are searching for these sprawling, wild, perennial herbs. Both have small, purple-brown, bad-smelling flowers, the

better to attract pollinators such as flies and carrion beetles. Both are the host plants of your larvae who eat the
leaves and absorb the aristolochic acids that make them
unpalatable to birds. Slowly, you fly above the grasses,
herbs, and shrubs, looking for the leaf shape you like best.

Today you like broad ovals. Last time you laid eggs, you
used a broad oval leaf, and now broad oval leaves are all
you can think about. They are your idée fixe.

In fact, it is March, and you are just beginning to lay
eggs, and the broad oval leaves of Texas Dutchman's pipe
are more abundant than the narrow leaves of Virginia
snakeroot. Later in the season, you may switch to a different shape. Your willingness to change is good, since the
broad leaves of Dutchman's pipe grow tough as they age,
become hard to digest, poor in nitrogen. Your larvae could
starve on such a plant. Virginia snakeroot, however, is always tender, even in May.

If you were capable of love, you would love your larvae
dearly, with all your long, tiny, kinked heart, with all the in-
stincts of your seed-sized brain. At the same time, your
particular caterpillars are not the easiest ones in the world
to care for. It's not their fault. You're not blaming them.

You blame the host plant. It's too small. Your voracious
eating machines eat it up and then have to eat a new one.
Sometimes a larva has to find and consume fifty plants be-

fore it can successfully pupate. The first plant, the plant you are searching for now, is the most important. The larger your larva is when it leaves that plant, the greater chance it has to survive.

Patiently, you fly above the grasses, herbs, and shrubs, looking for the leaf shape you like best and smelling for the odor of your host plant with your antenna, with the top of your head, with parts of your wings. Over and over, you land on a broad oval leaf. You tap the leaf with your foot.

You are proud of your feet, too, which have taste organs that can zero in on sugar as quickly as a third grader. When your feet taste something sweet, you lower your proboscis, ready to feed.

When you tap or drum a leaf with your foreleg, you are testing for the chemicals in your host plant. Because you are female, your forelegs are designed to support repeated strong strokes. On each foreleg is a cluster of sensory hairs that look like a tiny toothbrush. Spines on the tips of your forelegs may also pierce the leaf's skin so that juices and odors can flow to these hairs.

You drum, and in less than half a second, you know that this broad oval leaf is not the right broad oval leaf, not a Texas Dutchman's pipe but some other plant. You don't feel surprised. This is what usually happens. You spend most of your life on the wrong leaf.

You fly off. You land. You drum.

You fly off. You land. You drum.

You keep drumming.

Yes. Yes, this tastes right! This smells right. This is your larvae's host plant.

You stop drumming and look around. You are searching for eggs. If there are too many caterpillars on one plant, your larvae won't have enough food. They might start eating each other.

Now you see them: four reddish brown humps. In a matter of hours, tiny purple-brown caterpillars will break out of these humps. They will eat their shells and begin to eat whatever else they can find, feeding together in a group until they are larger.

You fly off in disgust.

You land. You drum. You fly off. You land. You drum. You fly off. You land. You drum.

You keep drumming. You stop. You look for reddish brown humps. This time, there are none. This time, the plant is perfect.

You fly off again.

You are a Pipevine Swallowtail, and it is still a mystery why and when you finally decide to lay your eggs.

You have a relative in Japan, another swallowtail, who determines the density of her host plant by landing on a certain number of them in a certain time frame. Her clutch size of eggs varies depending on that information. If there are a lot of host plants in the area, she will lay a lot of eggs; if there are not many host plants, she will lay fewer eggs.

You may be doing this, or you may not.

You are, in any event, taking your time, trying to be selective, and you feel smug about that. You know skippers who let their eggs drop haphazardly so that they fall anywhere about the plant. You know of Marbled Whites and Ringlets who broadcast their eggs while flying!

You don't pretend to be the perfect mother. For example, your larvae would survive better in a shady habitat than in a sunny one. But you still look for host plants in the sun and lay your eggs there. In this way, you avoid being eaten by orb-weaving spiders. Also, you need heat for your flight muscles to reach the right temperature, heat to generate energy to fly and search and fly and drum. (In another climate, you might worry more that fungal infections might kill your eggs, and you would reject any place too damp or dismal.)

At times, you stop your search to sip nectar from a flower. You are attracted to the colors pink and purple. You bask by spreading your wings and letting them radiate heat into the air under and around your body. You avoid or reject males who want to mate.

You fly. You land. You drum. You fly. You land. You drum.

Finally, on a Texas Dutchman's pipe free of eggs, you throw caution to the wind. *"Qué será, será,"* you hum impetuously.

You are weighty, filled with eggs, and you are ready to lay this burden down.

You curve your abdomen toward the leaf and position your ovipositor, the rounded end of a tube through which the eggs travel singly and are fertilized by sperm stored in your spermatophore. Your ovipositor has taste and sensory hairs that evaluate the site one more time. It may have simple eyes that further guide you.

You're like a driver backing her car out of a tight spot. You're like a pilot landing her aircraft on a naval ship. Careful, careful. Over here. Over here. Down. Put her down.

You have another important decision to make. You usually lay from two to five eggs. If this plant has a lot of leaves, if it seems young and healthy, you may tend toward the larger number. You may lay more eggs, too, if you haven't laid any in a long time, if you are feeling, perhaps, an uncomfortable urgency.

You decide to lay three eggs and you secrete a glue that attaches them to the underside of a leaf. It takes you about a minute. You fly off, and you don't look back. You've done your part. You will lay hundreds of eggs before you die. How many of these survive, if any survive, will depend largely on the site you have chosen for them.

Pipevine Swallowtail ovipositing

You have tried hard to balance the selection of that site with your own needs and with the pressure of time.

It hasn't been easy.

You're not complaining.

You don't expect any thanks.

NINE

ON THE MOVE

IN LATE SEPTEMBER 1921, FOR EVERY MINUTE of eighteen days, an estimated 25 million Snout butterflies passed over a 248-mile front from San Marcos, Texas, south to the Rio Grande. The flight may have involved 6 billion insects.

The Snout butterfly has a long appendage on the front of its head; when the adult rests parallel to a twig, its closed wings look like a leaf and its extended "nose," slightly angled, a leaf stem. In Texas, the upper wings of Snouts are brown or orange-brown with white spots.

The effect is Pinocchian.

In May 1978, Larry Gilbert, an entomologist at the University of Texas in Austin, was in a good position to observe and analyze the following:

A winter and spring drought in southern Texas had eliminated most of the parasitoid wasps that feed on Snout larvae. Good rains in May and June caused the butterfly population to explode in the next two or three generations. In July, the tropical storm Amelia produced even more rain. The larvae's host plant, the desert hackberry, flourished. The Snout caterpillars, green and dotted with yellow, flourished. The caterpillars defoliated their host plants and pupated. A population of a half billion butterflies emerged, and the Snouts began to migrate.

This is what happens when the world is too nice to butterflies.

The majority of the migrating insects were young males. Snout females are mated as soon as they emerge from their chrysalides, but at the original site, older males already waited by each pupa. So the younger males left, hoping to find unmated, available females elsewhere.

Some females also joined the migration, looking for a better place to lay their eggs. Most, however, stayed behind to exploit the dramatic response of the damaged hackberry—which was to sprout, as if it were spring, generous new leafy buds. Later, when the eggs hatched and the larvae began to feed on these leaves, many hackberries would die, preventing for some time future outbreaks of Snout butterflies.

Meanwhile, millions of Snouts filled the sky, their long noses pointed into the breeze. They clogged car radiators. They ruined laundry. They passed overhead like a muddy, aerial river.

When Larry Gilbert was a boy in south Texas, he also saw huge swarms of Snouts passing through after the summer rains followed drought, congregating in the hundreds on overripe dates in his grandmother's yard. These migrations paralleled a burst of green leaves, flowers, and other insects. "The sounds and fragrances of life were everywhere," he remembers, "where just days before, everything had been hot, dry, and bleak."

In 1977, Gilbert had access to a large protected reserve and was able to find the source of the river of butterflies that crossed a highway several miles away. Thousands of green and tan pupae hung from the leafless desert hackberry. The air shimmered with that connection between childhood and the rest of your life. The physics of time, once again, proved exotic.

"I confess I do not believe in time," Vladimir Nabokov wrote in his autobiography,

I like to fold my magic carpet, after use, in such a way as to superimpose one part of the pattern upon another. Let visitors trip. And the highest enjoyment of timelessness— in a landscape selected at random—is when I stand among rare butterflies and their food plants. This is ecstasy, and behind the ecstasy is something else, which is hard to explain. It is like a momentary vacuum into which rushes all that I love. A sense of oneness with sun and stone.

"You can imagine how pleased I was," Larry Gilbert says more mundanely, more modestly, "to satisfy my curiosity about the whys and wheres of Snout migratory flights."

Butterflies on the move, migrating butterflies, may set off as individuals, or in pairs; as small groups, or in large groups. Most regular migrants live in areas of great seasonal differences, summer and winter, wet and dry. They follow a seasonal pattern. A few butterflies follow vegetation, host plants and nectaring plants, up and down mountains. Some migrations, like that of the Snouts, happen irregularly, caused by an explosion of population, overcrowding, and competition.

Butterflies on the move are most often noticed when they move all together. Large numbers get our attention.

We like the abundance. A half billion. Six billion. We like to be overwhelmed, that Paleolithic thrill (without the danger) of being human in a world not dominated by humans.

The Painted Lady is the world's most common and abundant butterfly. Painted Ladies cannot survive extreme cold, so they often migrate south in winter, and north in spring and summer, their numbers estimated in the hundreds of millions, from Africa to Finland, from Mexico to Canada. In the summer of 1879, a migration of Painted Ladies through Europe was so great that it was called an invasion, one of nature's odd military campaigns.

A few years later, an explorer recorded the start of a Painted Lady migration in a stretch of grass on the Sudanese Red Sea coast:

From my camel, I noticed that the whole mass of grass seemed violently agitated, although there was no wind. On dismounting, I found that the motion was caused by the contortions of pupae of *V. cardui,* which were so numerous that almost every blade of grass seemed to bear one. The effect of these wrigglings was most peculiar—as if each grass stem was shaken separately, as indeed was the case. . . . Presently the pupae began to burst and the red fluid fell like a rain of blood. Myriads of butterflies, limp and helpless, sprinkled the ground. Presently the sun shone forth, and the insects began to dry their wings, and about half an hour after birth of the first, the whole swarm rose as a dense cloud and flew away eastward towards the sea.

The larvae of Painted Ladies migrate, too. In 1947, in the

Painted Ladies heading south

Saudi Arabian desert, a researcher observed an army of these caterpillars advance with the young hoppers of desert locust, eating the new spring growth.

In 1991, in California, a good year for butterfly eggs prompted another movement that entomologists could stand around and watch. By late May, hungry and over-crowded larvae had begun to crawl in straight lines in search of food. Compared to caterpillars raised alone, they showed increased activity, nervousness, cannibalism, and simultaneous pupation. The adults that emerged were also more active and gregarious. These adults had undeveloped reproductive organs and large reserves of fat. They didn't mate. They flew north instead.

It is possible that food shortages and overcrowding alter the behavior and biology of Painted Lady larvae, who de-velop into adults that quickly migrate. When the migrating adults start to feed, their hormone levels rise, and they then reproduce. Later generations of these Painted Ladies may experience cold weather; these butterflies, too, will begin to move.

In the first case, numbers cause migration. In the sec-ond case, temperature is the prompt.

We watch the spectacle. We enjoy the abundance. We indulge in metaphor. We want armies and clouds. We want whites like snowflakes, sulphurs like buttered popcorn. My own greed seems palpable. My eyes gleam. *A billion Snouts. Six billion.*

But I am a child of my time, and I do not see much excess in nature. Passenger pigeons once darkened the sky. Caribou stretched horizon to horizon. Salmon were so thick you could walk across water. This is not the coin of the twenty-first century. We measure our wealth by different standards.

I have seen three things:

The Apache del Bosque in central New Mexico is a flyway for migrating water birds, which come every winter. As the sun rises, tens of thousands of birds rise from the artificial lake, choreographed, wheeling and crying, honking and hooting, before they leave to feed on nearby fields. I have a photograph of my nine-year-old daughter watching this scene, the horizon pink, her pigtails longer than they will ever be again. The sky is a map pinpointed with ducks, geese, coots, cranes, and terns.

The river that flows by my home in New Mexico goes dry in the summer. Early in our marriage, years before we had children, my husband and I watched a large pool shrink with every hour. The water contained an abundance of tadpoles waiting to die as the pool evaporated. The animals squirmed against each other in a carpet of flesh. My husband and I watched, fascinated. I remember it still. This is what happens when you cannot move.

I remember the Monarchs, too, the ones that drowse every fall in eucalyptus groves near the Pacific Coast. On a dirt path, my sister chased my niece with a coat, for it was cold, surprisingly cold. The mentholated trees fluttered, covered with butterflies. We looked up. We spoke in whispers. The church of the Monarchs.

Each time, I felt rich, strangely buoyed.

Monarchs are the most famous migrating insect. Millions of these butterflies in Canada and the northern United States fly over 2,000 miles to overwinter in certain mountains in Mexico. (Some 5 percent of Monarchs, on the west side of the Continental Divide, fly to the Pacific Coast.) In the spring, these same individual butterflies start the return trip north.

Like Painted Ladies, Monarchs cannot survive an extreme winter. The Monarchs heading for Mexico take months to reach a roosting site that is generally above freezing but still cold enough to keep their metabolism and energy needs low for semihibernation. At this site, they have trees on which they can cluster. They have protection from wind and snow. They have water nearby. On warm days, they may rouse a little, fly a little, drink a little, and return to their somnolence, clinging to fir branches and to each other.

In March, they wake up. They feel the urge to mate. They move down the mountain and begin to fly north and east, on the lookout for milkweed plants. Before they die, females lay their eggs, recolonizing the southern United States. The generation that emerges now will continue to fly north, to mate and lay eggs, and to die in about a month. The next generation will do the same, and the next, until the last generation of butterflies reaches the most northern edge where Monarchs and milkweed plants can live.

By the end of summer, the world is colored in Monarchs again. The orange-and-black wings gladden the human heart. Entomologists smile more often. Children are happier.

The Monarchs that now pupate and emerge in late summer and early fall will be different from those of previous generations. In the larva and in the chrysalis, shorter days and cooler temperatures have triggered hormonal changes. The adult males and females are reproductively immature. At the first sign of cold weather, they stir collectively, a shiver of desire, and begin moving south—to a land they have never known.

These butterflies have unusually long life spans, as long as nine months, time for them to fly to their overwintering grounds, to dream through the winter, to mate in the spring, and to start the journey back north. Unlike their parents, they are not solitary but huddle together, roosting at night as they hurry south. During the day, they fly in clouds, as high as 1,000 feet and as fast and far as fifty miles. As they go, they stop and feed. They even gain weight.

Somehow, too, these individuals know where to go. They follow a map not in our dimension for certain mountains in Mexico, certain south-facing slopes, certain fir and pine trees.

For navigation, they use the sun. In one experiment in the American Midwest, researcher Sandra Perez netted a random group of migrating Monarchs and kept them in her lab for two weeks. There she changed their light and dark cycle so that the butterflies became accustomed to a different time zone. When the Monarchs were released, one by one, they flew in the wrong direction, based on where the sun should have been, based on what time they thought it was.

On cloudy days, the Monarch relies on a magnetic compass: tiny bits of magnetite in the thorax. When Sandra and her colleagues exposed migrating fall Monarchs to normal magnetic fields, the butterflies flew normally toward the southwest, on to Mexico. When the butterflies were exposed to a reversed magnetic field, they flew in the reverse direction, northeast. When they were exposed to nothing, when they had no magnetic field, they flew hither and thither.

Monarchs, like other migrants, probably also use visual landmarks. Butterflies commonly compensate for cross-wind drift; in experiments across open bodies of water, butterflies able to see landmarks on the horizon compensated better. Sulphurs and skippers flying over ocean areas without landmarks also seemed to use, with moderate success, nonstationary objects such as clouds or rippling waves.

Most butterflies get through life by flapping their wings, the basic flight pattern, up, down, up, down. But a Canadian Monarch needs to reach its Mexican wintering site in less than ten weeks. The most earnest flapping is not good enough. Instead, Monarchs use thermal updrafts to glide and soar like eagles. In the late afternoon, they may stop migrating for the day because the earth has cooled and the thermals have gone. Monarchs also take advantage of the wind if it is blowing their way. If it is not, they adjust by flying closer to the ground, where there is less wind. They may choose to rest, eat, drink, and wait.

Other migrating butterflies have different patterns. Many fly in a straight line, purposeful and low. One scientist noted that "even when a migrating butterfly is trapped on a porch, it appears to be trying to batter down the house, and will persist in a chosen course rather than retreat a few yards or deviate from it." Cloudless Sulphurs, Gulf Fritillaries, and Long-tailed Skippers all tend to travel in the less windy "boundary layer," a few yards above the surface of the earth. Great Southern Whites, moving up and down the coast of Florida, fly in streams forty-five feet wide, rarely more than twelve feet above the ground. On breezy days, they use sand dunes as a buffer. On still days, they fly directly over the dunes. Solitary Red Admirals, migrating from northern Europe to a warmer place in Spain, are often seen at about waist level, set on a determined track. Put something in that butterfly's way and it will swerve around, or over, reorient, and continue, straight ahead.

The Monarch is at one end of a scale. At the other end is the Colorado Hairstreak, which commonly strays a few yards in its lifetime. The stability of the host plant may help determine how much a butterfly will travel. Stay-at-home butterflies usually lay their eggs on dependable perennials such as trees. Peripatetic butterflies tend to lay eggs on less dependable plants such as weeds and annuals.

Within a species, individuals vary. In a population of migrants, some stay put. Equally, a nonmigrant may find itself on the road for good cause.

Butterflies everywhere are on the move, moving toward heat, away from cold, moving toward food, away from scarcity, moving to find a mate, a nicer neighborhood, more opportunity.

Pack your bag, don't think twice.

What happens to butterflies that don't migrate?

In the winter, they may hibernate. Some species hibernate as eggs, some as larvae. Many hibernate as pupae, some as adults. A few hibernate one year in one stage and one year in another.

In hibernation, everything slows down. No one hatches. No one molts. No one pupates. No one mates. No one produces eggs.

Larvae find a good hiding place, under a leaf, in the grass, in your garden. Adults find a good hiding place, in a tree, under a leaf, in your garage. Occasionally, they may fly out and feed.

The blood thickens with an antifreezing agent that acts like glycerol. The water content of the system decreases. Free water converts to a gelatinous colloid.

In conditions of heat or drought, butterflies aestivate, which is the same idea.

Stop

moving.

Butterflies everywhere are on the move, and butterflies are

not alone. Each year, freshwater eels slither through dewy grass to reach the ocean. Seabirds clock in 20,000 miles. Walrus cover a tenth of that. Mexican free-tail bats fly across the desert. Microscopic flatworms wiggle eight inches, twice a day. In one month, in April 2002, a group watching migrations tracked whooping cranes, humpback whales, Monarchs, hummingbirds, caribou, bald eagles, and robins.

They missed the flatworms.

I like the numbers, the big numbers. More is better. More butterflies are better than fewer butterflies. A river of butterflies is a wonderful thing. Millions of butterflies are the jackpot. I like the largesse, the almost casual gesture, as if a generous earth were whispering into my ear, "See how I replenish myself, see how I birth and birth and birth and darken the skies and fill the waters and cover the ground and still I have more to give."

IN THE LAND OF BUTTERFLIES

*H*IS DESTINY WAS TO MAKE AND SELL STOCK-ings. His maternal grandfather had been a wool-stapler, his paternal grandfather a master dyer of hosiery. His father rose to the height of owning a workshop. In 1838, when Henry Walter Bates was fourteen years old, his formal schooling came to an end and he was apprenticed to a wholesaler, a business in which he worked thirteen hours a day, six days a week, arriving at seven to open and sweep the warehouse, closing at eight when darkness shut down the streets of Leicester, the center of English hosiery manufacturing.

It was the time of the Industrial Revolution, of polluted cities and dirty skies, when mechanized factories replaced the small workshops of people like Bates's father. The combination of man and machine seemed stronger than any force of nature, and almost immediately we mourned our victory. Some of our best nature writing comes from this period in the mid-nineteenth century, the "golden age" of natural history collecting, the heyday of the butterfly enthusiast.

As amateur naturalists explored the countryside, they killed and pinned great numbers of insects, many of them previously unknown to science. Social clubs, created for the working and clerical classes, held self-improvement courses and conducted field trips. On his days off, and through the summer evenings, Henry Bates learned to draw, translated Homer's *Odyssey,* and began a collection of beetles. At the age of eighteen, he published his first article, "Note on Coleopterous Insects Frequenting Damp Places," in *The Zoologist.* At nineteen, he met fellow enthusiast Alfred Russel Wallace, who taught English at the local Leicester Collegiate School. Twenty-one years old, six feet and two inches tall, Wallace was sturdy and gangly while Bates was slight and frail, suffering from chronic indigestion, acne, and poor circulation.

The two got along. They admired each other's insect collections. They read the same books: Darwin's *Voyage of the Beagle* and Malthus's *On the Principle of Population.* They worried like puppies over the most exciting scientific mystery of their time: How, why, and when did different species arise in the world's array of creatures? They clung together like damned souls on a greased sled hurtling toward a conventional English life of mercantile ambition and minimal advance up the social ladder. They each acquired a hundred pounds and determined to launch a private collecting trip to some more exotic, sun-filled, blue-skied land.

At the natural history department of the British Museum, an entomologist suggested Brazil. The museum itself offered to buy any rare insects and birds they collected. An agent was found to handle their specimens.

And they were gone, in April 1848, aboard the trading ship *Mischief*.

From the beginning, in an early letter to Bates, Wallace had proposed that they were setting out not on a young men's lark but with the greater goal of "solving the problem of the origin of the species." (In his study in England, still musing on his experiences in the Galapagos Islands, Charles Darwin pursued the same goal.)

After a swift passage from the Irish Channel to the equator, Bates and Wallace arrived at a small Brazilian village near the Pará River. In his best-selling travel adventure *The Naturalist on the River Amazon*, Bates would later remember, "It was with deep interest that my companion and myself, both now about to see and examine the beauties of a tropical country for the first time, gazed on the land where I, at least, eventually spent eleven of the best years of my life."

The two would stay a year and a half on the Pará, taking odd trips, collecting, and exploring. Of butterflies, Bates noted with satisfaction that while the British Isles had only 66 species, and all of Europe only 321, he could find 700 within an hour's walk from the town where he now lived.

He marveled at the swallowtails, "so conspicuous in their velvety-black, green, and rose-coloured hues," which flew lazily about the streets and gardens, often entering through windows to investigate a house. The splendid blue metallic morphos measured seven inches across, flapping huge wings like "birds on the verandah." In the dry season, especially, the collecting was rich, and "an infinite number of curious and rare species may then be taken, most diversified in habits, mode of flight, colours, and markings; some yellow, others bright red, green, purple, and blue, and many bordered or spangled with metallic lines and spots of a silvery or golden lustre."

A clearwing, its wings transparent except for a spot of violet or rose, looked like "a wandering petal" as it flew low over dead leaves in the gloomy shade. Congregations of yellow and orange butterflies packed together on the beach, their wings upright, so that the sand seemed "variegated with beds of crocuses."

Heliconius

Then there was the genus *Heliconius,* the longwings and postmans, their wings a deep black decorated with streaks of red, white, orange, and yellow. Their shape was elegant and aristocratic, as was their flight, slow and languid. Their numbers amazed the young naturalist, all those "bright flakes of colour" sailing through the air.

A later writer, from another time and place, would echo the sentiment: "We're not in Kansas anymore."

Wallace and Bates parted company to collect on different parts of the river system. Bates headed for the upper Amazon. In a letter to his brother, he wrote of a typical collecting day, and of his typical attire, when between 9:00 and 10:00 A.M., he headed for the woods wearing a colored shirt, trousers, boots, and an old felt hat. Over his left shoulder, he carried a double-barreled shotgun, used mainly to shoot down birds and animals. In his right hand, he carried a net for insects. On his left side, he slung a leather bag with a pocket for his insect box and a pocket for powder and shot. On his right side dangled a bag for game, "an ornamental affair with red leather trappings and thongs to hang lizards, snakes, frogs, or large birds." One pocket in this bag contained his percussion caps, another held papers for wrapping up delicate specimens, another for "wads, cotton, box of powdered plaster, and a box with damp cork for micro-lepidoptera." On his shirt, he had a pincushion with six sizes of pins.

In just a few minutes, after entering the trees, the collector arrived "at the heart of the wilderness." Before him was nothing but forest. "Many butterflies are found. I walk about a mile straight ahead lingering in rich spots, and diverging often. It is generally near two p.m. when I reach home, thoroughly tired. I get dinner, lay in hammock a while reading, then commence preparing my captures, etc: this generally takes me to five p.m.; in the evening I take tea, write and read but am generally in bed by nine."

Unlike most collectors, Bates often settled in an area.

He lived for over four years near the town of Ega, long enough to be a welcome guest at weddings and the christenings of babies. Never robust, he suffered from parasites, yellow fever, malaria, and dysentery. His diet was poor and low in protein. He made friends, yet he was often lonely, and in "these savage solitudes," he missed, above all, books and conversation.

At one point, he came into the possession of two children. The boy later became his assistant in the field and then a goldsmith. The younger girl was ill and died, despite considerable nursing. "She was always smiling and full of talk," the naturalist wrote sadly. "It was inexpressibly touching to hear her, as she lay, repeating by the hour the verses she had been taught to recite with her companions in her native village."

To the annoyance of the other Europeans in Ega, a grieving Bates had the child baptized and buried in a robe of calico, her hands over her breast, a crown of flowers in her hair.

Despite the joy of collecting butterflies, and his dread of returning to the "slavery of English mercantile life," Bates admitted to his brother, "I was obliged, at last, to come to the conclusion that the contemplation of Nature alone is not sufficient to fill the human heart and mind."

In the meantime, Alfred Wallace had left the Amazon after only four years. On his return trip to England, his ship caught fire, and all his collections were lost. Undeterred, the young man went out again, this time to the Malayan Archipelago. In 1855, from a mountain hut in Sarawak, he polished up a paper called "On the Law Which Has Regulated the Introduction of New Species." In 1858, a joint reading of two papers before the Linnean Society, one written by Wallace and one by Darwin, outlined the theory of natural selection and its role in the evolution of species. A year later, Darwin's long-researched *Origin of Species* appeared in print.

That same year, Bates also left South America, never to return. In the last decade, he had collected and transported more than 14,000 different insect species, 8,000 of these previously unknown. Moreover, he had made observations of mimicry among *Heliconius* butterflies that would be of lasting importance.

On his final night on the Pará River, he endured vivid and unexpected memories of England:

> Pictures of startling clearness rose up of the gloomy winters, the long gray twilights, murky atmosphere, elongated shadows, chilly springs, and sloppy summers; of factory chimneys and crowds of grimy operatives rung to work in early morning by factory bells; of union workhouses, confined rooms, artificial cares, and slavish conventionalities.

To live again amidst these dull scenes I was quitting a country of perpetual summer.

One senses some ambivalence.

The following day, his ship drifted out of the mouth of the Pará. He reached out his hand toward the land of butterflies, and that "was the last I saw of the Great River."

Life righted itself soon enough. Bates went to live with his parents in Leicester and returned to the hosiery business. He continued work on his collection of species, selling and classifying them. He and Darwin began a correspondence, congratulating each other on their work and consoling each other for their bad health.

In 1861, Bates read his own paper about Amazonian butterflies before the Linnean Society. Deliberately, he yoked his ideas to the service of Darwin and Wallace. The naturalist understood that he was in a unique position: He had made his observations for a very long time, very carefully, in the field. What he saw was a range of butterfly species of many races and geographical variations. Among these were parent species side by side with daughter species that had evolved from the parents into something new. What he saw, in living flakes of color, was proof of the mutability of species.

More important, Bates had been struck by butterflies such as the black, orange, and yellow *Dismorphia,* which mimicked members of the toxic *Heliconius.* The imitation, Bates deduced, was for the purpose of protection: The mimic was trying to look like the bad-tasting model. Not only did butterflies in other families mimic *Heliconius,* but some *Heliconius* "are themselves the imitators; in other words, they counterfeit each other, and this to a considerable extent."

Bates's recognition of mimicry and his ideas about its causes and effects have been further developed, but not supplanted. A hundred and fifty years later, his work still stands.

Today, we distinguish between Batesian mimicry, in which a good-tasting species mimics a bad-tasting one, and Mullerian mimicry (named for Fritz Muller, another Amazon explorer), in which bad-tasting models mimic each other.

In Batesian mimicry, the mimic is a parasite, since the model gains nothing and even loses some protection when predators occasionally taste the mimic and suffer no ill effect. Batesian mimicry tends to evolve with a major genetic change in the mimic's pattern, causing it to resemble its model weakly, followed by more gradual refinements.

The model, meanwhile, will try to evolve away from the

mimic. Too many good-tasting mimics and the predator wises up, at which point the model is likely to be sampled, too.

At the same time, most bad-tasting butterflies are tough, their bodies able to withstand a tentative bite. It is the tasty mimic who gets tested and then eaten. The learning abilities of birds and other predators and their willingness to try out their prey constantly affect the evolution of a model/mimic system.

In Mullerian mimicry, all the bad-tasting mimics gain as they reinforce the predator's understanding not to eat any of them. Fewer of any one species get tested. Everyone wins.

Mullerian mimicry tends to evolve with the least bad-tasting butterfly resembling the most bad-tasting one; then all the butterflies start to look like each other.

Butterflies can alter their patterns relatively easily because each separate wing part develops on its own. Killing a small group of cells in a pupating Buckeye will stop the formation of the large eyespot on its forewing, but other design elements on the wing are not affected. In the "nymphalid ground plan," a butterfly can change its spots, one spot at a time.

Both Batesian and Mullerian mimics often combine to form a mimicry ring, a group of species, good-tasting and bad-tasting, that share a similar pattern.

Efficiently, it would seem to make sense for butterflies to agree on one universal warning sign—a big red circle with a diagonal slash.

In fact, there are many different mimicry rings, each one with a distinct pattern. The tropical *Heliconius* combine with other butterfly species, as well as some day-flying moths, to form the "striped tiger" ring, the "red" ring, the "blue" ring, the "transparent" ring, and the "orange" ring. Each of these can include up to two dozen species of different families and subfamilies. These different patterns may come about according to where the species of the ring evolved, where the butterflies roost at night, where they fly during the day, or how they select mates. Geographical boundaries or barriers also affect the evolution of patterns.

Lepidopterists continue to marvel over the *Heliconius,* whose species readily change patterns to adapt to geographical conditions. For example, some twelve different races, or variations, of both the Postman and the Small Postman exist across South and Central America. The two species mimic each other, and unless you are examining them very closely, they are indistinguishable. Their co-mimicry is consistent from region to region, race to race, so that a Postman will look quite unlike its own species in another area and exactly like a local but different species. (Almost always, one species practices pupal mating and one does not. Different mating tactics may allow the two similar-looking but distinct species to occupy the same small habitat.)

Mimicry becomes more complex when gender-based morphs of a species are involved. In some mimicry rings, the male of a species is a nonmimic, but as many as four different forms of the female mimic other bad-tasting species, thus joining four different mimicry rings. (Males may have a resistance toward changing their color pattern because they use color as an important signal among themselves.)

The world of mimicry rings is a dizzying one. Like hula hoops, the rings seem to go faster and faster. They intertwine like Celtic knots. They induce a certain vertigo.

The North American ring of swallowtails might begin with the bad-tasting Pipevine Swallowtail, who has absorbed aristolochic acids as a larva from plants such as Texas Dutchman's pipe and Virginia snakeroot. The Black Swallowtail is its Mullerian mimic or co-model; it resembles the Pipevine and also tastes bad, having eaten toxins from the parsley and carrots in your garden. The Spicebush Swallowtail is another co-model, full of benzoic acid. The Queen Swallowtail, on the other hand, is a good-tasting Batesian mimic. The dark morph female of the yellow Tiger Swallowtail is a second good-tasting Batesian mimic; these dark forms occur only in areas where the Pipevine

Swallowtail coexists. The Red-spotted Purple, from another butterfly family, is a good-tasting Batesian mimic of the Pipevine and also occurs only in southern areas where that swallowtail lives. The northern form of the Red-spotted Purple is called the White Admiral, whose broad white bands make it look quite different. Finally, the female of the Diana Fritillary is a blue-black mimic of the Pipevine Swallowtail, while the male Diana Fritillary is orange and brown-black. No one knows whether these female fritillaries are Batesian or Mullerian.

Mimicry rings spin faster, more crazily, when you remember that the unpalatability of a butterfly is on a spectrum. The Black Swallowtail is not half as nasty as the Pipevine Swallowtail. Related species vary, as can members within a species. Season or weather affects the amount and quality of poisons in a host plant, which then affects individual butterflies, as well as populations.

Researchers often get confused, as in the famous case of the Viceroy butterfly. For many years, we thought it was a Batesian mimic, a parasite of the Monarch and the Queen. We know now that birds spit out Viceroys, too. This species is also bad-tasting, a Mullerian mimic.

At the same time, a Monarch raised on certain host plants may not be poisonous at all and so becomes a Batesian "automimic," copying its own species.

Larry Gilbert theorizes that the *Heliconius* genus first evolved as a Batesian mimic of other toxic butterflies. This protection gave the adults more time to spend on flowers and allowed them to develop pollen feeding, which enhanced their ability to synthesize and produce the poisons that would make them distasteful to predators, turning them into Mullerians.

Butterflies mimic not only wing pattern but flight pattern too the wing-beat frequencies and asymmetrical wing motions commonly found in toxic species. Bad-tasting butterflies tend to flap and sail slowly. Their slender bodies are another visual signal, and they often have long abdomens that may enhance balance and make for a smoother flight.

Good-tasting species tend to fly fast and irregularly. For obvious reasons, they skip, dart, and change course quickly. Their larger flight muscles may require a wide thorax and a thick, chunky body. Their abdomens tend to be short and hidden by their hindwings, harder to grab, harder to find, and perhaps easier to maneuver.

Batesian and Mullerian mimicry are distinct but malleable strategies. Mullerian mimics switch to Batesian and Batesian to Mullerian. Both systems are vulnerable to the changing palatability of prey and the changing responses of predators. These things happen in real time, under complex conditions.

Some birds, for example, have learned to gut Monarchs and eat only the nontoxic parts, like a Japanese chef preparing a blowfish.

What does this mean to a passing Viceroy?

Lecturing his audience in 1861, Henry Bates sternly emphasized that scientists had to leave their laboratories for "the workshop of nature." Perhaps only a man who had spent eleven years in the field, dug parasites from his feet, shivered with yellow fever, and buried children in the jungle could truly understand.

Concerning the mimicry of *Heliconius* butterflies, he cautioned, "The faithfulness of the resemblance, in many cases, is not so striking when they are seen in the cabinet. Although I had daily practice in insect-collecting for many years, and was always on my guard, I was constantly being deceived when in the woods."

And he concluded that his observations offered "the most beautiful proof of the truth of the theory of natural selection," the "selecting agent being insectivorous animals" who implacably destroy unsuccessful mimics.

Soon after this lecture, Bates was encouraged by Darwin to begin a travel memoir, published successfully in 1863. He retired from the family hosiery business and married, at age thirty-seven, the twenty-two-year-old daughter of a Leicester butcher. The couple bought a small house on the outskirts of London, where the naturalist hoped to continue his research.

Instead, he found himself drudging for a living, editing books and cataloguing private collections. Not even the British Museum would give him a position. For years, the self-taught explorer would be snubbed by academics and the scientific establishment, if not by Darwin and other powerful figures. As one of these wrote consolingly: "Entomologists are a poor set, and it behooves you to remember that in dealing with them. It is their misfortune, not their fault."

Eventually, Bates accepted the role of assistant secretary of the Royal Geographical Society, a job in which he met most of the famous adventurers of his day, men such as Stanley Livingston and Richard Burton. He ended up with a large family of sons and daughters, and he continued to write and edit, to receive increasing recognition, and to become the world's authority on ground beetles.

He died in 1892, at the age of sixty-six.

Alfred Russel Wallace also went on to marry and raise a family in England and to struggle eternally with money and bills. In history, he would become the better known of the two: the co-discoverer of the theory of evolution.

Looking back, in 1906, Wallace could recall only the pertinent parts of his first meeting with his lifelong friend:

"How I was introduced to Henry Walter Bates? I do not exactly remember, but I rather think I heard him mentioned as an enthusiastic entomologist, and met him at the library. I found that his specialty was beetle collecting, though he also had a good set of butterflies."

ELEVEN

THE NATURAL HISTORY MUSEUM

THE NATURAL HISTORY MUSEUM IN LONDON, which holds one of the world's largest and oldest butterfly collections, looks like a cathedral. The entrance is massive, patterned in fawn and blue-gray stone, with rows of arched windows and rising pinnacles. Inside, stained glass spills light onto the central nave, or "sacred space," which opens into smaller "chapel areas." The high ceiling is decorated with green and gold illustrations of plants identified by their scientific names: *Digitalis purpurea, Rosa canina, Daphne laureola*. The sacred space is filled with the replicated skeleton of the dinosaur *Diplodocus*, the bones and the air between bones of a plant-eater who measured eighty-five feet from head to tail. The chapel areas showcase fossils, the mystery of what has been and will never be again.

This church crawls with animals, molded and cast in

terra-cotta on leafy panels on every wall: warthogs and owls, foxes and sheep, pigeons and stoats. Snakes entwine pillars. Monkeys climb doors. A lizard creeps toward the exit sign. The west half of the building highlights living species, the east half-extinct ones.

The museum's marriage of religion and biology began with its first donor, Sir Hans Sloane, whose private collection formed the basis of the British Museum in the 1750s and later the splintered-off British Natural History Museum in 1881. Sloane's will expressed the hope that the study of his natural oddities would result in the higher study of God. The first superintendent of the natural history department was Richard Owen, the man who coined the word *dinosaur,* a creationist in fierce opposition to Darwin's theory of evolution. Owen also believed that the museum's purpose was to display Divine Will. With the help of the brilliant architect Alfred Waterhouse, he created a building designed for worship.

For years, a bronze statue of Sir Richard Owen glared across the central hall at a marble Charles Darwin sitting in repose. Today, Darwin looks out virtually unnoticed over a café area serving lattes and fruit tarts. His anonymity seems less banishment than acknowledgment. He's common as cake, in the air we breathe and food we eat.

Dick Vane-Wright is the museum's Keeper of Entomology, and his personal secretary is at the reception desk, ready to take me via lifts and locked doors to the hidden kingdom. First we go through the dinosaur exhibit, turning a corner to confront the most realistic robotic *Tyrannosaurus rex* I have ever seen.

As big as my bedroom, he crouches over the body of a duck-billed *Edmontosaurus* twitching between life and death. *T. rex* growls and swings his head from side to side, moving red-stained teeth closer to the low wall that separates us. Bending to sniff his prey, he roars loudly, then rears back with nervous excitement. He is about to rip the flesh from the dying herbivore. Yet, he does not. There is a beat. *T. rex* growls and swings his head from side to side, moving red-stained teeth closer . . .

London is in the middle of a rail crisis, and Dick Vane-Wright is running late. I am taken, in the meantime, to see Jeremy Holloway, a scientific associate. Jeremy's office is on the ground floor of a department that goes up and down six floors and contains some 30 million insects in 120,000 drawers. Of those, 8.5 million are pinned moths and butterflies. These include the over 2 million specimens that Lord Walter Rothschild gave the museum in 1937, which were combined with the earlier collections of Sir Hans Sloane and James Petiver. Other collections have been donated or bought, and the museum's associates, including

Jeremy, and John Tennent in the Solomon Islands, continue to add specimens, perhaps as many as a thousand a year.

Jeremy is actually working on moths. He may know more about the larger moths of Southeast Asia than anyone alive, and is currently writing the definitive *Moths of Borneo*. He stands up to open a drawer. His office is small, more of a carved-out space than an office, on the edge of a cavernous floor area with rows and rows of tall wooden cabinets. The paths between the rows are like the corridors of a maze. The cabinets, many of which were donated by Lord Rothschild, rise above our heads and then spread laterally to a distant wall. Jeremy's collected specimens, pinned directly into the drawer, are nicely at hand.

And there they are: a line of delicate brown-and-white moths all looking somewhat alike. Jeremy points out the printed label that says "type." The type specimen is the individual moth or butterfly that has been chosen to represent the species as a whole. These are the dead diplomats of their countries. The Natural History Museum has over half the type specimens of the world's moths and butterflies. The rest of the series of somewhat look-alike insects shows the variations within a species, comparative to the type, including geographic races.

Because of storage problems, collectors no longer pin

the kind of long series they used to delight in. This may be unfortunate. Many of the insects that Jeremy describes in *Moths of Borneo* are new species. They have been collected before, but when Jeremy looks at these older collections (specifically at the genitalia, an important diagnostic key), he finds in a series of moths not one, but two or more different species.

A large collection such as this is used mainly for taxonomy. In the Solomon Islands, John Tennent swoops down his net to catch a blue butterfly so that it can be compared with all the other blues here, including those swooped by A. S. Meek and other early collectors. These comparisons help determine the butterfly's family tree, its affinities, and its relationships.

A second use of the collection has to do with biodiversity. This museum is a window onto the butterflies that could have been found in England three hundred years ago, when James Petiver was receiving specimens from Eleanor Glanville; or one hundred years ago, when Walter Rothschild was a boy collecting in his estate gardens outside London; or fifty years ago, when Miriam Rothschild was collecting in the same gardens. In these drawers, we see what we have lost and what we have not.

Jeremy is interested in how moth and butterfly diversity can reflect overall diversity. He believes there is a correlation. He is looking at which moth species exist in different managed landscapes: in a new softwood timber plantation in Malaysia, in an older plantation with a thick understory, and in a primary undisturbed forest.

"When we know what is there," Jeremy says, "we can go back and keep checking against that record. We can see what is changing. We can see the costs in biodiversity of a certain practice, and maybe how to mitigate those costs. First we need some basic information."

In a world that has become a garden, with all our landscapes managed, determining the "right mosaic of development" may be as important as conserving patches of wild land.

Moths of Borneo is an eighteen-volume series. Jeremy is on the thirteenth book.

Dick Vane-Wright, Keeper of Entomology, has an office with a view that I fancy includes the treetops of Kensington Garden, made famous in James M. Barrie's *Peter Pan*. In 1961, when Dick was eighteen years old, he came to this office to see about a job. Except for leaves to get his degrees and to do research, he has worked for the museum ever since.

John Tennent has urged me to ask Dick about the time he ate insects on television.

Philip DeVries has described Dick as an "unusual and ebullient" classifier of things who, like Phil, is also a former jazz musician. During our talk, the keeper will jump up every so often to fetch this book or that one down from the shelf. He jumps up now to show me the 1885 tract "Why Not Eat Insects?"

"Why not eat insects? Why not indeed!" begins that author, going on to discuss the nutritional, culinary, and economic value of sawflies and wood lice. Concerning the green worm of the Cabbage White butterfly, "I see every reason why cabbages should be served up surrounded by a delicately flavored fringe of the caterpillars which feed upon them!"

The plump bodies of moths are nice when grilled.

And "Let us, then, cast aside our foolish prejudice, and delight in chrysalides fried in butter, with yolk of egg and seasoning, or 'Chrysalides à la Chinoise.'"

Dick Vane-Wright is almost as cheery. "Eating insects is a challenge of social mores and cultural norms. It's puncturing people's pomposity!"

Dick confides, "There is a certain pomp and pomposity to *this* job, which goes against my grain."

So when the Natural History Museum decided to reprint the 1885 tract, Dick went on a tour of promotion, crunching locusts over the radio and frying up mealworms on the BBC. For a while, a few London restaurants tried serving insect food. "I had a modest part in that," the Keeper of Entomology says now.

Ask entomologists how they became interested in bugs and butterflies and three out of four will give you a number.

"I was thirteen years old," writes Edward O. Wilson.

"Although I lacked an inquiring mind, I was a good observer for my five years," begins Miriam Rothschild.

"When I was twelve years old," Robert Pyle, a well-known lepidopterist, remembers.

"I was a child on country walks," Jeremy Holloway agrees.

"I was a little boy collecting beetles," David Carter, Jeremy's colleague, will say.

"I was seven years old," Dick chimes in, "and I was given a book by some children's author now much reviled, soft information on natural history, with colored pictures of butterflies. Out in our yard we had a flowering tree and, in my mind now, it seems I was able to match every butterfly in that book to a butterfly on that tree. It was a one-to-one match. *Very* satisfying."

Matching things one-to-one, recognizing patterns, finding the thing that does not fit the pattern, naming the pattern, naming the anomaly, checking back with the book, writing the book yourself someday. It is all, indeed, very satisfying. The adult murmurs of biological order. The seven-year-old shouts, "Gotcha!" This surpasses the swoop of a net. In taxonomy, you capture a species.

Biological order today is based on how organisms have evolved through time: the ancestors they share. Dick likes to spend his time unraveling what the colors on a swallow-tail or the perfume of a Monarch might reveal about its ancestry and closest relatives. As part of his research in the 1980s, he went to the Philippines, an area rich in butterflies, but already known to be badly deforested.

"I wasn't prepared for how deforested," Dick says.

It made a huge impact on the rest of my life. I was so shocked and so convinced that this wasn't good for anyone, certainly not for the people who lived there. It was a mixture of greed, ignorance, and poverty, destroying an ecosystem and replacing it with nothing at all. I became ill. I thought I was physically ill but, as it turned out, I was mentally or emotionally ill. The problem was fixed when I had to fly to New Guinea to do more experiments on butterfly behavior. The landscape there was magical. I felt immediately better. I had simply, actually, been depressed.

Dick jumps up to look for a book on New Guinea butterflies. He finds, instead, something produced by a Japanese collector. What is happening now to butterflies in Japan?

No one knows.

"Unless we have an idea of what is there and where it is and how to identify it, we can't take care of it," Dick echoes Jeremy.

The London museum is now mapping the hot spots of diversity. "In effect, we're telling people that if you must trash x amount, please don't trash this part because it will have tremendous impact on how many species can remain in the world."

As fast as possible, lists of species are being generated: for Asia, for Africa, for Australia, for North America. Dick calls this bioaccounting.

"We're producing telephone directories."

What happens when you call a number in this directory? Not much if you are looking for information on how a butterfly lives, mates, reproduces, or dies. Don't expect a chatty conversation.

"There is so *much* we don't know!" Dick says, sounding excited and distressed at the same time. "You could spend a week studying some obscure insect and you would then know more than anyone else on the planet. Our ignorance is profound."

He jumps up to fetch another book.

In 1984, Phil Ackery, a collections manager at the museum, and Dick Vane-Wright collaborated on two important books, *Milkweed Butterflies, Their Cladistics and Biology* and *The Biology of Butterflies*. By 1990, Phil was feeling the need to specialize, to focus on research or on collection management, but not on both. "I landed rather more on the collection side of the fence," he says today.

Much of his work deals with pest control.

Collectors have long complained about the problems of maintenance. In 1702, Eleanor Glanville wrote to James Petiver:

> I being not at home have preserved but few plants this year, and so long neglecting to clean my butterflys being almost 2 years ye mites have done me much mischefe, I have lost above a 100 Species of my finest . . . wch I put up closest and Safest for fear of Spiders and mice. I believe for want of aire, not being fresh, ye mites breed ye more and ye Bettles was molded over with a whit crusty mould wch when I went to clean broke al to pccccs. I hope while I live never again to let them be so long neglected.

The Natural History Museum in London is mostly plagued by the larvae of beetles: *Anthrenus sarnicus,* a gray-and-gold carpet beetle about a tenth of an inch long; *Reesa vespulae,* the American wasp beetle, the females of which can reproduce on their own; *Attagenus smirnovi,* the brown carpet beetle; and *Stegobium paniceum,* the biscuit, or drugstore, beetle. Phil and Dick refer to them generically as museum beetles, of which there is also a specific species, *Anthrenus museorum.*

Typically, a female beetle living in one of the building's outside birds' nests flies through a window to lay her eggs near something that smells good, perhaps a dead insect on the other side of a wooden cabinet. The larvae hatch, worm through the tiniest crack, and begin to feed. A new kind of museum beetle, introduced in the last twenty years, has the destructive pattern of taking a bite from one butterfly and then moving on to take a bite from another, potentially plowing through half the specimens in a case. By now, the beetle itself is too large to escape. A researcher opens a drawer. One dead beetle. Lots of little bits of butterflies.

Phil guesses that at the beginning of each year there might be forty identified infestations in the museum's 120,000 insect drawers. For a long time, the museum relied on pesticides, until people realized that what was toxic to invertebrates was toxic to mammals, too. In this case, the chemicals were contained under high vapor pressure. When a researcher opened a drawer, the fumes roiled out. Fifteen years later, as I walk past the cabinets, fifteen years after the last dose of Naphthalene, I can smell it still.

Collections now have to survive without insecticide. First, everything is frozen, whatever people send in, whatever the museum buys, at minus thirty degrees Centigrade for seventy-two hours. This kills pests like beetles. Infested drawers also go into deep freeze. Slowly, too, Phil is acquiring "good museum furniture," not the antique glow of wood but gray steel boxes with tight seals. These boxes are compacted into unaesthetic horizontal units that can be opened manually to a specific site.

Eventually the entire building will have to be hermetically sealed, and people will have to change their work habits. Phil points to a sink in his own slightly cluttered office. "That creates a micro-environment of high dampness. There's lots of book lice running around a sink like that. Someone comes and places a drawer by the sink and there you are."

Some of Phil's clutter involves his most recent project, an exhibit of seventy-three butterfly species collected by Henry Walter Bates. Phil was interested in how Bates had pinned his butterflies and how he had transported them to the British Museum.

Would I like to see?

I am out of the office in a nanosecond, possibly carrying a few book lice with me.

Phil takes the lead through the wooden canyons. He has worked at the museum since 1965 and has written amusingly of those annual meetings in the 1960s and 1970s, attended by prominent lepidopterists, when "the canyons between the cabinets would echo with such triumphant cries as 'New record for Shropshire!' always in the well-modulated but irritatingly penetrating tones of what seem to be the primary attribute of a British private education."

"How did I get interested?" he repeats my question.

Well, I wasn't one of those people born with a net in my hand. I wasn't rearing Large Whites when I was three and a half years old. I suppose that for most people working in a natural history museum, the attraction is to the natural history, whereas mine was for the museum part. I quite enjoy ordering things into straight lines and then putting pretty labels on them. It wouldn't make odds to me whether it was butterflies or flintlock rifles. I'm happy doing things like indexing. I have a good boredom threshold.

We stop before a cabinet. Phil casually opens a drawer. They are so much brighter than I had imagined they would be. The blue-and-yellow *Nessaea batesii*. The azure of *Asterope sapphira*.

Phil believes that Bates set and pinned his specimens in the field and that later, after he had sent them to his agent, they were repinned in another style, slightly damaging the thorax.

Phil shows me a damaged thorax.

Without my having to beg, he takes me to look at the Ornithoptera, the Queen Alexandra's Birdwing and the Giant Birdwing, also collected in the nineteenth century. Just a few of these green-and-blue wings fill up the space of a drawer. The yellow abdomens are huge. The green is primal.

Phil points to holes in the wings of certain females. In 1890, one collector wrote of seeing such a birdwing while bathing; he scrambled from the water, seized his net, and ran through the jungle naked: "I tread upon a sharp stone and fall head over heels, but picking myself up again, continue the chase along the beach, till at last, just as my quarry is rising among the trees, I come up with it, and by a well-directed stroke enclose it in the net. I leave it to any ardent entomologist to imagine my feelings on this occasion."

Later, the happy nudist "saw several more but as they kept high up among the trees I thought I would try to shoot

Queen Alexandra's Birdwing

them with dust-shot. I was carrying a 16-bore gun, into one barrel of which a Morris tube .360 bore was fitted, and by its aid I shot two more females."

Phil closes the drawer, rolling the birdwings back into darkness, like jewelry hidden away. He has to return to work, and he tells me that I can wait for David Carter, another collections manager, at a table here, at the end of this corridor, under the canyon walls, not far from the butterflies collected by Henry Bates. It's been fun, Phil says. Ta-ta.

I can't believe they trust me, alone.

I sit at the table, waiting for David Carter.

Then I stand up and sneak to the nearest drawer.

I open it slowly, trying not to make a sound. Slowly, I reveal rows of spotted green and spiraled red, creamy forewings, chevroned hindwings. I leave the drawer open and move to another.

Camberwell Beauties. The gorgeous Peacock. The Tortoiseshell.

I tiptoe down the wooden canyon, open two more drawers, three more, five more, an Owl, a Zebra Longwing, a Red Admiral. I leave them all open. The butterflies begin to stir, pushing their wings against the case, moving up, bright ghosts, through the glass into the air.

I open more drawers, and more. The room fills with butterflies, series of Checkered Whites and blues, Tiger Swallowtails, clearwings, metalmarks, Snouts, Ornithoptera. One Grayling bows to another. Two sulphurs begin to mate.

There are tropical cries, parrots and monkeys. There is the scent of jasmine.

And then I am back at the table, looking innocent.

David Carter finds me waiting. We talk about how old brass pins can suddenly implode an insect's body in a reaction of temperature and pressure and fat. He tells me about creaking cabinets and the sounds they make at night when he is working here alone. Sometimes the wood splits like a pistol shot! We talk about research and a recent request from scientists who wanted to run DNA tests of butterflies collected long ago. They wondered if the museum "could spare a leg."

We wander into the collection, up and down stairs, and I confess, "I still don't know where I am going here," and David agrees, "Oh, it was years before I knew where I was going."

Then he is using his keys to open up a gray steel cabinet. This is the collection from Sir Hans Sloane, who in the early 1700s bought James Petiver's collection. Sir Hans was so appalled by Petiver's "bowerbird's mentality," the shelves of objects helter-skelter, that he immediately hired someone to conserve the specimens. I am looking now at three-hundred-year-old sulphurs pressed between sheets of thin clear mica. The sulphurs still glow with the incandescent color that gave them their name.

Somewhere here is the fritillary that Eleanor Glanville sent her friend and mentor, after her second husband left her in peace, before he kidnapped her son and threatened her other children, before she began to confuse her children with fairies. The Glanville Fritillary, never common, remains rare, confined to a small area off the south coast of England.

And there is more, even before Petiver, a collector who pressed his insects, like flowers, between the pages of a book. David Carter, already enthused, is sparkling now, generating excitement. This may be the oldest insect collection anywhere.

Surely David has done this before, many times, first showing the visitor photographs of what is inside the book, a pressed Tortoiseshell, a Peacock, and then showing the book itself. But, of course, not opening it. Every time the volume is opened, more damage is done to its fragile contents. Of course, we cannot open the book. We can only gaze at its cover.

Surely David has done this before, yet his interest seems as keen and fresh as these butterflies, remarkably still alive in their drawers. It is David's obsession, and Dick's and Phil's, to keep this collection keen and fresh for another three hundred years. It is a remarkable continuum and a singular statement that goes beyond naming, beyond ownership, beyond order. It is the shape of history: stories and time.

A man runs naked through the jungle. A man leaps, ardent, a net in his hand.

TWELVE

NOT A
BUTTERFLY

CONCEALED AMONG THE FLOWERS, THE GOLD-enrod Stowaway flutters up, a dab of butter in the sun. Its shiny yellow wings are streaked with orange. It is not a butterfly.

The Grapevine Epimenis is black with a large red patch on its hindwing and a large white patch on its forewing. During the day, it feeds on wild grapes in the dappled woods of eastern North America. It is commonly mistaken for a butterfly.

The upper green forewings of the Scarlet Tiger moth are speckled in yellow. Its hindwings are matador red.

A moth in India is laced with patterns of green, black, orange, and white, overlaid with a blue metallic sheen.

One day-flying moth looks like a swallowtail.

Another shimmers like a rainbow.

What is the difference between a moth and a butterfly?

Entomologists find the question tiresome. According to their nature, they look defensive or chagrined. There is often too little difference, and scientists understand how unscientific this must seem.

Of some 165,000 species of Lepidoptera, we have decided that about 11 percent are butterflies. The rest are moths, and the majority of these are micromoths, or microlepidoptera, usually small and primitive in the sense that they evolved first, before butterflies. From 50 to 100 million years ago, butterflies and a few other families of moths, called the macromoths, or macrolepidoptera, developed from this original group.

The two superfamilies of butterflies, the Papilionoidea and Hesperiodea, have distinct traits separate from most macromoths.

Notably, most butterflies are active during the day. They rely on sight to find food, host plants, and each other, and they use visual signals, design and color, to communicate with friends and enemies.

Some researchers believe that butterflies moved into the sun to escape predation by bats; that, in effect, bats invented butterflies.

Certainly, bats helped shape moths. Bats emit ultrasonic cries and use echolocation to zero in on night-flying insects. In response, night-flying moths tend to be furry, which may obscure their radar profile. Some moths also developed "ears" on the wings, thorax, and abdomen that are sensitive to ultrasonic sound. Hearing a bat close by, the moth dive-bombs to the ground. Some moths produce their own ultrasonic squeaks and clicking noises that may confuse the bat's radar system. Just as likely, these sounds warn the bat that the moth is toxic, an audio form of the Monarch's wings.

Spiders also prey on moths, fishing with their webs as the insects careen blindly in the dark. Moths and butterflies escape spider silk by shedding their wing scales (which detach easily) and slipping free. Spiders have learned to distinguish the vibration of a fluttering moth from that of a fly or a bee, and they rush immediately to bite the former before it can escape.

Some spiders build their webs in columns, towers of silk, catching and recatching the moth as it flutters free and up, free and up, until its wing scales are gone, and the bald wings are easily caught and held.

Flying at night means that moths rely heavily on scent to find food and mates. Spiders take advantage of this, too, by sending out a lure of fake sex pheromones. Male moths hurry to the bait, where they are caught with strands of specially prepared superstrength glue.

As a group, butterflies exchanged these night dangers for

the new dangers of daylight, a world of birds with excellent eyesight and color vision. Some moth species in different families chose to do the same thing, and so we have a bright day-flying moth in the same family as a dull night-flying cousin.

Antennae also distinguish butterflies from moths. A butterfly's antenna will end in a thickened knob, or club. A moth's antenna may taper to a point, or look saw-toothed, or resemble a feather or a palm frond. Antennae are mainly used for smell, and moths are champion smellers. Moths win all the smelling bees. From experiments in the lab, we know that male hawkmoths can smell and distinguish almost every compound we throw at them. We know that the male silkmoth, with its huge plumose antennae, can detect the sexual attractant of a female silkmoth at volumes of only a thousand *molecules* per cubic centimeter. We know that some male moths can scent and track a female from over a mile away.

In the dark world of moths, females do most of the calling for mates, sending out their chemical bouquet from a gland on the abdomen. According to her species, the female calls at certain times under certain conditions in certain places. The males are poised to receive the message, their antennae sweeping and filtering the air. The male smells the call, follows the odor plume, finds the female, and emits his own chemical signal. With the female already in charge, courtship is usually quick and easy. So is copulation.

A third way to distinguish butterflies from moths is to look for a tiny bit of carpentry in the structure of the wings. Butterflies lack the catch-and-bristle arrangement, common to most moths, that hooks the fore- and hindwing together. In flight, this helps the wings beat together as a unit.

Also, butterflies tend to rest with their wings closed over their backs. They fly or bask in the sun with wings spread out horizontally. Moths tend to rest with their wings either folded up in an angled tent or flat and horizontal like a basking butterfly.

Eggs and larvae have their own peculiarities: the position of a pore, a special gland in the neck, various tufts of hair.

Exceptions abound. Skipper butterflies can be drab and small, fold their wings in a tentlike position, and have barely thickened antennae. Burnet moths are red-spotted day-fliers whose antennae seem noticeably clublike.

One group of "butterfly-moths" is such a mix that taxonomists recently added them to the butterfly family. The neotropical Hedyloidae have ears on their wings and are mostly dull-colored and small. They include day-flying and night-flying species. They do not have clubbed antennae, but, like swallowtails, they spin girdles and their eggs and caterpillars seem very butterflyish.

Another group of larger tropical "butterfly-moths" fly mostly during the day, are boldly colored, and have clubbed antennae; their caterpillars are distinctly moth-like.

For now, they are not butterflies.

Count up the mammal, bird, reptile, amphibian, and fish species. Add them together. There are still more moths. In such a large group, you are guaranteed a range of adaptation.

You can anticipate the fun.

Some moths are so small that they spend their entire larval lives mining the inner cells of a leaf. The tunnels of these leaf miners create characteristic designs: a delicate spiral, a simple maze.

Other larvae excavate tree trunks, feeding lugubriously on wood pulp for as long as four years and emitting strong-smelling frass in great quantity from their burrows.

There are moth larvae that live in ponds, eat pondweed, make shelters from aquatic leaves, and use feathery tracheal gills to draw oxygen from the water.

There are moth larvae that build silk "bags," which they

carry about and camouflage with debris and evergreen nee-
dles. As an adult, the male escapes from his shelter. The
adult female does not. Even after metamorphosis, she
lacks legs or wings or eyes and is not much more than a
sack of eggs waiting to be found and fertilized.

The caterpillars of a moth in Arizona feed on the small
flowers of oak trees, which they mimic with yellow-green
skin and fake pollen sacs. Later in summer, after the flowers are gone, the next generation of larvae look like oak
twigs, with bigger and heavier jaws for eating leaves. Scientists once thought these were two species. They are, instead, the different costumes of one.

The largest moth, from South America, has a wingspan
of a foot.

A hawkmoth in Madagascar uncurls a foot-long proboscis to fit into the foot-long nectary tube of the orchid it
pollinates.

A moth in Asia pierces skin and sucks blood.

A luna moth has no mouth.

The ascetic Yucca moth also does not eat or drink but
pollinates the yucca flower by collecting pollen, flying to
another yucca plant, and depositing her load on the waiting stigma. At that point, the female lays her eggs in the
flower's ovary. The flower becomes a pod full of seeds.
When the larvae hatch, they consume a percentage of

these, burrow out, fall to the ground, and pupate. The Yucca moth is one of the few insects that pollinates actively, deliberately, as a way to ensure food for her young.

Hornet moths are a caricature of what they imitate, with the long, translucent wings of a wasp and a fat yellow-and-black-striped abdomen. These moths buzz viciously and pump out their stomachs as though about to sting.

Other moths resemble bumblebees.

Some hover like hummingbirds.

A moth in Venezuela imitates a cockroach.

In their great range and number, moths have a greater influence than butterflies. They are better and more important pollinators of flowers and crops. Their caterpillars feed the world. We have even domesticated moths, as we have sheep, turning them into small silk factories. We wear their excretions proudly.

Moths are bigger pests, too. They eat flour and clothes. They devour crops and gardens. The Gypsy moth has defoliated forests.

Our cultural associations with moths tend to the negative. Like butterflies, they represent the souls of the dead, but their visitations are less benign. Moths bring bad luck. They foretell mischief. They come from the shadows. They are hairy and gray. They fly grotesquely, suicidally, into candles and lamps and porch lights. (Their compound eyes

Death's-head Hawk moth

may actually be seeing a very dark area next to the bright light; they are trying to fly into that dark zone.)

Think of the Death's-head Hawk moth. Patterned in yellow and black, it weighs as much as a mouse and has the design of a skull on its upper back. Its name, *Acherontia atropos,* comes from the Greek *Acheron,* the river of pain in the underworld, and *atropos,* one of the three Fates who cut the thread of life. The moth squeaks when disturbed and uses its short pointed proboscis to pierce through the wax of beehives, from which it steals honey. The skull mark may mimic a queen bee's face, so that bee workers will not attack the intruder. The moth's squeak may further pacify the insects.

In the movie *The Silence of the Lambs,* a serial killer breeds Death's-head Hawk moths and places their pupae in the throats of his victims.

In a fifteenth-century manuscript, a Death's-head Hawk moth is painted in the corner of the page dedicated to Saint Vincent, who represents immortality, the triumph over death.

Moths represent the dying that comes before eternal life, the gloomier side of resurrection.

Give them their due. Moths are beautiful. Moths are complex.

But they are not butterflies.

THIRTEEN

TIMELINE

The El Segundo Blue spends most of its life on the minute flower heads of coast buckwheat. From mid-June to mid-August, the female lays fifteen to twenty eggs a day, which hatch in five to seven days. The larvae are highly, remarkably varied, from pure white to dull yellow, from red to maroon, patterned in yellow or white dashes and chevrons. The caterpillars feed on and hide in the tiny petals, stamens, stigma, seeds, and leaves of their host plant. In their third instar, they develop honey glands and are tended by ants, who protect them from parasitic wasps and other predators. A single caterpillar will eat two or three flower heads in eighteen to twenty-five days before it crawls or drops to the ground, burrows two inches deep into the debris of its home buckwheat, and pupates through the fall and winter.

The adult El Segundo Blue emerges when the coast buckwheat is flowering again. The butterfly has the radius of a dime. The male's upper wings are a shimmery, silvery blue with black-and-orange borders edged in white. The female's upper wings are brown with an orange border. The female flies immediately to a flower head to await a patrolling male who finds and mates with her in a matter of hours, punctual as a well-run bus. In the wild, she will live from two to seven days, nectaring continuously, egg laying continuously, and trying to avoid the lynx and crab spiders that live in one out of two hundred flower heads. In a laboratory, raised with tenderness by scientist Rudi Mattoni, she would live an average of sixteen days.

The female blue shares her flower heads with a hairstreak butterfly, the Acmon Blue butterfly, and at least eight moth species. All these larvae are sometimes cannibalistic. They compete for food, and they harbor the parasitoids that never rest but breed and move year-round from host to host.

Other beetles, flies, crickets, weevils, and gnats tend to their own business with the coast buckwheat; the plant, in turn, has a profound relationship with the soil and shifting sand, as well as subtle ties, good and bad, to neighboring primrose, deerweed, sunflower, lupine, and bladderpod, which together support lizards, toads, mice, shrews, foxes, and owls.

No one can say how long this has been going on, this

being the intimate, familial life of the El Segundo Blue and the coast buckwheat with their multiple forms of kinship and strife. But one could guess that the cycle has lasted for thousands of years, repeating like a beloved soap opera on an eight-mile stretch of the El Segundo dune system in glamorous southern California.

In the fifteenth century, as they walked the shoreline looking for food, Native Americans brushed their hands over coast buckwheat and startled up small blue wings. After the Spanish conquered these tribes, conquistadors and priests stood on the dunes; and then, after the Mexican Revolution, independent mestizos; and then, after the United States won its war with Mexico, American settlers—a parade of human beings always conquering and being conquered, always feeding from the land.

By the 1880s, ranchers had long grazed cattle, horses, and sheep on the coastal prairie east of the dunes, and farmers were replacing native vegetation with beans and corn. The small communities of Redondo Beach and Venice crept up onto the sand itself. In 1911, an oil company built a refinery above the beach.

In 1927, Rudi Mattoni was born in Venice, California. He spent most of his childhood a few miles north in Beverly Hills, where he became a "boy collector," much like Richard Vane-Wright, Vladimir Nabokov, and countless

El Segundo Blue

others. It was a sport not so different from the rural tradition of hunting and fishing: finding and catching and owning something beautiful. At that time, butterflies were still abundant in Los Angeles, and from his bedroom window, Rudi could lean out and collect six different species from one bush.

In 1927, too, a plane piloted by Charles Lindbergh and humorist Will Rogers landed on a dirt runway east of the El Segundo dunes. The site was eventually chosen as the city's new airport.

The 1929 stock market crash and the Great Depression slowed the development of the dunes until after World War II, when there was an explosion of workers who needed housing.

By the 1950s, a subdivision covered much of the El Segundo Blue habitat, right under the flight path of jet planes leaving an increasingly busy LAX airport. Residents complained about the noise. The Federal Aviation Authority worried about public safety. Meanwhile, the city of Los Angeles was buying up much of the surrounding land.

In 1957, Rudi Mattoni earned his Ph.D. in zoology and genetics from the University of California at Los Angeles. He went to work determining the effects of the first atomic bomb explosion on insect populations in New Mexico. He would later do research for the U.S. space program, looking at the microecology of long flights: the genetics and population dynamics of bacteria in hypogravity and irradiation. He would also teach, study the Sonora Blue butterfly, develop new methods in the commercial farming of mushrooms, standardize protocols for hundreds of tests on agricultural products, and help produce

3 million sterilized cotton pink bollworms for biological control of that pest.

In 1965, a dispute between policemen and an African American community precipitated the Watts riot in South Central Los Angeles; thirty-four died and more than 1,000 were injured. Parts of the city burned to the ground, never to be rebuilt.

From 1966 to 1972, the conflict between residents on the El Segundo dunes and the LAX airport was resolved. Over eight hundred houses were purchased or condemned and then bulldozed.

In 1971, Arthur Bonner was born, the third of five children. His family would soon move from rural Florida to South Central Los Angeles, where two rival African American street gangs, the Crips and the Bloods, were at the height of a turf war.

In 1973, the president of the United States signed into law the Endangered Species Act (ESA), the world's only legal prohibition against the extinction of other species, even those as small and localized as the El Segundo Blue butterfly.

In 1975, to realign a major highway, a large area of the remaining El Segundo dunes was excavated, recontoured, and stabilized with native seed. Unfortunately, the seeds were native to a coastal sage, not a dune scrub plant community.

The revegetation effort introduced the highly successful common buckwheat, which is toxic to the larvae of the El Segundo Blue. Moreover, the common buckwheat blooms a month earlier than coast buckwheat, providing food for the blue's competitors. These two butterfly and eight moth species can produce multiple generations in a year. More butterflies and more moths also meant more parasitoids.

Also in 1975, thanks to members of the conservation group, the Xerxes Society, Standard Oil Company agreed to fence off and manage their small portion of the El Segundo Blue habitat. This was the first formal butterfly reserve in California.

In 1976, the El Segundo Blue was listed as protected under the Endangered Species Act. In the early 1980s, the butterfly had a population of about 1,500 on the one-and-two-thirds-acre site at the oil refinery, and of about 400 on patches of scrub dune vegetation on the three hundred acres south of the LAX airport, land that still contained the rubble of the condemned subdivision. Under the ESA, these acres were proposed as critical habitat for the blue.

A few people had a better idea: a twenty-seven-hole golf course.

In 1982, in the sixth grade, Arthur Bonner helped out a Crip gang member during a fight by firing a gun into the air. The Crips nicknamed the boy Bub.

In 1983, the golf-course proposal was submitted by the Los Angeles City Planning Department to the Coastal Commission, the organization overseeing development on California's shoreline. By now, there had been eight public hearings with all sides vocally represented, some for full development of the remaining dunes, some for partial development, some for zero development.

As one chronicler noted, the El Segundo Blue "became a convenient rallying point." The butterfly was a marker of spiritual growth. The butterfly was one of multiple tests to determine whether human beings would survive and deserved to survive our unthinking, rapacious, relentless desires. The butterfly stood for humanity's relationship with the natural world.

In 1983, Rudi Mattoni and others declared a nearby species, the Palos Verdes Blue, officially extinct. Its habitat was twelve miles from the El Segundo dunes. For the last few years, Rudi had been counting the number of Palos Verdes adults he could find on two hands: six, four, seven, zero.

In 1984, Arthur Bonner dropped out of school to sell drugs and steal cars.

In 1985, the Coastal Commission rejected the plan promoted by airport officials for a twenty-seven-hole golf course with eighty acres set aside as a preserve for the El

Segundo Blue. Instead, the airport was directed to protect and study the butterfly, starting immediately. The Board of Airport Commissioners gave Rudi Mattoni a small grant to stabilize the population and to begin a biological survey of the three-hundred-acre area.

That study would identify eleven new plant and animal species unique to the dunes and threatened by competition with non-native species. These included the El Segundo Giant Flower-loving Fly, the San Diego Horned Lizard, the El Segundo Spineflower, and the El Segundo Jerusalem Cricket.

"Bad News for Golfers" ran a local headline when Rudi went public with his conclusion that the site was not only a major habitat of the endangered El Segundo Blue but its *only* habitat, as well as being a "hotspot of diversity" for other species.

In 1988, on the day his first son was born, seventeen-year-old Arthur Bonner sat on a bus headed for state prison. He had shot a security guard in the face.

In 1989, a major restoration project began on the El Segundo dunes, powered by volunteers from a local conservation group, Rhapsody in Green. Every third Sunday, men and women from around Los Angeles came with trash bags, gloves, and earplugs; they dug up introduced and invading plants such as ripgut, iceplant, acacia, and common buckwheat, and replaced them with the El Segundo Blue's host plant, the coast buckwheat. On some days, Rudi could be seen directing them, wearing a pith helmet.

In a few years, parts of the dune looked, Rudi says now, "like they were supposed to look." The estimated number of El Segundo Blues rose to 3,000.

In 1991, the Los Angeles City Council voted that two hundred acres of the dune system be permanently preserved. Rudi was given $430,000 in state highway mitigation funds to direct the restoration effort.

In 1992, the Crips and Bloods negotiated a peace, based on a 1949 United Nations Middle East treaty.

In 1993, Arthur Bonner was released from jail. He had already decided that staying out of trouble and being a good father were his new priorities. His brother suggested he join the Los Angeles Conservation Corp, which paid inner-city youth minimum wage to work at projects in southern California. Arthur took the job and was sent to work clearing brush behind the LAX airport, where a chain-link fence enclosed a sandy area heralded by a sign: "The El Segundo Blue Butterfly Habitat."

The story goes that Arthur asked some guy next to him how anyone could keep a butterfly locked up with a chain-link fence. Rudi replied that the butterflies would stay near their food and host plants. He gave Arthur some books on the subject.

Soon enough, Arthur began coming as a volunteer, every third Sunday.

In 1994, Rudi began an insect survey on a small area owned by the United States Navy in the Palos Verdes peninsula. Nearby, men on bulldozers were replacing an underground pipeline. Rudi saw a little blue thing, whoosh, fly by. He caught it. In his hand he held a Palos Verdes Blue.

The caterpillars of the Palos Verdes Blue spend most of their life inside a milkvetch seedpod, where they feed on seeds high in protein and fat. They get into the pod by making a small hole, through which ants later enter to protect the larvae in exchange for a good time, a bit of honeydew, maybe a song or a pheromone. The caterpillars will also feed and live among the flower heads of deerweed. The adults emerge from late January through March and fly about for five days. The males, an inch across, have the typical shimmery blue wings edged in white; the females are brownish blue. The underside of both are light gray with black dots circled in white.

Rudi was off like a car alarm to the men working on bulldozers. "You'll have to stop, guys," he said, and remembers later, "They were quite nice about it." Then he called the

U.S. Fish and Wildlife Service. An extinct species had reappeared, some two hundred blues surviving among the scrub brush and storage tanks of a government fuel depot.

The Department of Defense, which controls 25 million acres in the United States, has a profound understanding of the Endangered Species Act. Some biologists consider our often huge military bases to be inadvertent "arks," protected to an extraordinary degree from public access and public use (such as grazing). These bases are home to over one hundred threatened or endangered species.

The navy responded promptly. They worked with Rudi to monitor the butterfly's population. They built a small lab on the site that could breed Palos Verdes Blues to be released into the wild. And they began clearing non-native vegetation and reestablishing over thirty historic plant species, including milkvetch and deerweed.

"Revegetation is the key," Rudi says. "Solve the plant problem and you solve the butterfly problem."

Later that year, Rudi hired Arthur Bonner to work full-time at the new Palos Verdes lab. Eventually, Arthur would be responsible for the replanting of the coastal sage community and for the lab's rearing of captive butterflies.

In 1997, Rudi and Arthur each received a Special Conservation Achievement Award from the National Wildlife Federation.

In various accounts of his life, Arthur usually says a version of the same thing, "I'm saving these butterflies from extinction, and they're saving me, too."

"Of everything I have ever done in my career as a scientist, in all my jobs," Rudi Mattoni says, "this work on the dunes is the most important."

By 2003, estimates for the El Segundo Blues ranged from 15,000 to 50,000. Rudi worries that further restoration work at the butterfly habitat is not getting done. Non-native species are creeping back in. Still, both the El Segundo and Palos Verdes Blues now have a home.

Arthur Bonner continues to work at the Palos Verdes lab and to take inner-city kids out on field trips. They look at caterpillars. They look at butterflies. Some of these children need a lot of convincing. They find it hard to believe that the one thing turns into the other.

This is *your* home. This is Los Angeles, Arthur tells them. There are miracles like this every day.

FOURTEEN

THE BUSINESS
OF BUTTERFLIES

MIRIAM ROTHSCHILD WISELY WROTE, "YOU
must see for yourself the electric brilliance of *Morpho
cypris*, a drifting fragment of sunny sky, before you can be
deeply moved by the description of its undulating, glitter-
ing flight and fleeting azure presence. Gossip about
friends, we know, is delectable, but about strangers is bor-
ing beyond belief."

Philip DeVries says that a collector's trick is to wave a
blue silk scarf in the air when a morpho passes overhead.
A male will dive down to investigate; a female couldn't care
less. Presumably, the iridescent blue male is being territorial, upset by the presence of another male.

A bit of fluttering silk is a fair description for the blue
morpho, as well as a lure, though the collector may still
lose her specimen. A morpho in flight, so seemingly languid, can suddenly jerk up like a balloon yanked by its

string, or a leaf caught in the wind. For someone with a butterfly net, there is no second chance.

The first morpho I saw in Costa Rica was patrolling a river in the lowland rainforest next to the Caribbean Sea. It indeed looked like a piece of sky had detached itself and flapped away. I almost expected two wings, an outline of the escaped creature, to appear cut from the air.

In different species, the upper wings of a morpho range from blue to violet to white. Females are generally drabber. The underside is always cryptic, patterns of brown and cream. Eyespots on the underside can be another defense, as can the erratic flight. Otherwise, the morpho seems an exception to the rule that bold colors advertise toxicity; these butterflies, with their larvae and pupae, are readily eaten by birds.

The brilliance of any morpho is structural, a play of light, like the sky itself.

In a Costa Rican gift shop, I can buy a pinned morpho for $25. On the Internet, a pair of *Morpho cypris* goes for $119, plus shipping. A certain birdwing from Asia is $1,000. Worldwide, the sales volume for butterfly collecting is over $100 million.

On one Web site, the butterflies are organized by color, not species. I am assured that my morphos will contrast well with the included slate blue shadowbox frame.

The commercialization of butterflies is not necessarily a problem in the conservation of butterflies. Sometimes it can even be a help.

Tortuguero is a Costa Rican community of about six hundred people on the Caribbean coast. The village, once dependent on fishing and logging, is now surrounded by national parks and preserves. The main economy is tourism. Most people here work in lodges or drive the boats that transport visitors around a natural system of canals. There is no road into Tortuguero. There is only water.

The main economy in all of Costa Rica is tourism and, with 30 percent of its land protected from development, that industry can be called ecotourism. People visit Costa Rica to see the rainforest, the cloud forest, the volcanoes, and the beaches. In these landscapes, they want jaguars, monkeys, toucans, parrots, and morphos. They want biodiversity in the context of a hotel.

In the American discussion of how to use natural resources, a bumper sticker in my town reads, "You can't eat scenery." But in Costa Rica, that's what they do.

Turtles are a good example. The beaches near Tortuguero are a major nesting site for the endangered Loggerhead, Hawksbill, Leatherback, and Green Turtles, which the villagers once hunted for meat and eggs. Today, turtles are protected by these same villagers, who give nightly tours that let people watch a three-hundred-pound Leatherback dig her nest in the sand. Reservations at hotels and lodges are strongly recommended in the turtle nesting season.

Daryl Loth is a Canadian who came here nine years ago to manage a biology field station in the Barra del Colorado Wildlife Refuge. He married a Costa Rican and is now starting a bed-and-breakfast in Tortuguero. His two-year-old daughter sits on his lap. His wife and nine-day-old son are inside the house. The Rio Tortuguero flows a few feet away.

Behind Daryl's bed-and-breakfast, behind the Catholic church, to the right, past the store, is the butterfly house that aid agencies have been trying to establish in the village. The start-up money has built a small building with a yard fenced in by mesh. Host plants and nectar sources have been planted or brought inside the enclosed area.

The idea is to have major species, some morphos, some postmans, some swallowtails, flying about in a small space. Tourists will pay a few dollars to walk through a door and see butterflies so easily, so many butterflies, so close, flashing their eyespots, uncoiling their proboscides, feeding on rotten fruit.

There are over fifty larger but similar butterfly houses in the world, mostly in Europe and the United States. They are deservedly popular, for they deliver exactly on what they promise—detail, intimacy, the sensation of flying flowers. In a butterfly house, you are Henry Walter Bates. You may experience, as Alfred Wallace did on collecting a particular birdwing, "an intense excitement," your heart beating wildly and the blood rushing to your head. You may feel a slight headache the rest of the day.

In addition, the butterfly house in Tortuguero plans to breed butterflies to sell as pupae and to buy pupae from local breeders. This is called butterfly farming. Since butterflies live for only a short time, the fifty major butterfly houses in the world, often associated with zoos and natural history museums, need a constant fresh supply. Tortuguero would supply, specifically, the butterfly house at the Toronto Zoo, one of the supporters of this project. And that, as Daryl says, "makes for a nice circle."

In Tortuguero, the butterfly house and farm will not be engaged much in butterfly ranching, a system in which people who live on the edge of a wild area grow host and nectar plants in order to lure butterflies in from their natural habitat. Because the eggs and larvae are watched and protected, many more butterflies survive than usual. For ranchers, the wild area is now a resource of free-ranging butterflies, the place from which their revenue comes.

Butterfly ranching has been most successful in Papua New Guinea, where the government regulates both ranching and collecting: buying butterflies exclusively from Papua New Guinea villagers, keeping prices relatively high, and protecting threatened and endangered species. Each year, the government sells its butterflies to collectors, scientists, artists, and houses. Village-based butterfly ranching is now being started for popular commercial species such as the Paradise Birdwing.

It is not a coincidence, perhaps, that Papua New Guinea still has large tracts of virgin rainforest and the world's only constitution to name insects as a natural resource.

In Kenya, a butterfly ranching project can be directly linked to the conservation of the important preserve Arabuke-Sokoke. Ten years ago, 83 percent of the preserve's neighboring landowners wanted at least some of the forest cleared for logging and agriculture; over half wanted it cut down completely. These farmers were almost all poor, living on their own edge of survival. Today, with butterfly ranchers getting a dollar a pupa, the same people can earn more from wild swallowtails and fruit-feeding charaxes than from their combined crops of mangoes, coconuts, and cashews. A recent survey showed that only 16 percent of the local community now want the forest destroyed.

What Tortuguero's butterfly house may eventually become is a version of the Costa Rica Entomological Supply company (CRES) near the capital of San José. CRES is the largest exporter of butterfly pupae in Costa Rica and one of the largest in the world, sending out an average of 6,000 pupae a week to places everywhere, from the Budapest Zoo to the Houston Museum of Natural History. The company employs some sixty breeders, mostly family operations in the rural area around San José, who farm the insects in their backyards. A small percentage ranch. The breeders bring their product to a central location, where the pupae are slotted into an efficiently run business, the employees busy in the office receiving, sorting, and packing; checking lists of who wants what; checking for disease; checking for fungus; and double-checking the documents needed to transport insects across national borders. CRES also runs its own butterfly house, called The Butterfly Farm, which thousands of tourists visit every year.

Butterfly houses have specific requirements. The ecology and behavior of a species must be well-known, and the butterflies should be bright and showy and fly well, not crash into windows or try to ascend past ceilings for a vertical mating dance. Preferably, they should land on top of leaves instead of hiding under them. It is also nice if they live for a few weeks.

Breeders, too, whether farming or ranching, need species that have known host and nectar plants and that are relatively tough. Their caterpillars should not be too voracious, devouring a garden overnight, and their pupae should not be too sensitive to movement or humidity.

In Costa Rica, this leaves about sixty species suitable for export. Houses buy pupae for $2 to $4 each, and a typical breeder for CRES can make $500 a month, which is considered a good wage. A very industrious breeder can make five times that.

"Breeders aren't made, they're born," the owner of CRES says. "It takes extraordinary interest, a drive, a sensibility and a hard-work ethic. It's dawn to dusk. You have to be out there monitoring eggs, dealing with ants, watching for parasites and predators. Most of my breeders are self-taught. They have initiative. They have a passion for what they do."

The owner of CRES is an American, a former Peace Corps volunteer who began his business in 1991, inspired by E. Schumacher's book *Small Is Beautiful*. Some precepts for a beautiful business are that it must be sustainable, it must use appropriate technology, and it must be intellectually interesting.

The Tortuguero project fits into the beautiful paradigm, and Daryl Loth emphasizes how the project will also fit into village life, in small and intangible ways. Daryl foresees high school students running the butterfly house, learning English so they can talk with tourists, learning how to run a business, and learning some basic biology. Daryl is really more interested in community than in enterprise. He hopes to give local people "a few more options, pulling them out of a rut, helping some kids have a little more self-esteem." He is cautiously optimistic. After several false starts, including a recent fire, the butterfly house in Tortuguero is about to be revived.

Those of us who watch small children do it obliquely, not wanting to startle or intrude. We try to avoid being caught in a frank, fascinated stare. Daryl's two-year-old daughter has her hand in his shirt, the familiar gesture of a toddler who no longer nurses but still likes the comfort of touching a parent's chest. Forty-five babies were born in Tortuguero in 2001, and Daryl tells me about the school system, how they struggled to get the high school back into the village, how so few children will ever go past high school. Before we leave, I get to look at the new baby.

Like Henry Walter Bates, I have come to the conclusion that "the contemplation of Nature alone is not sufficient to fill the human heart and mind."

For this reason, I have brought with me my own seventeen-year-old daughter, who sits patiently through my conversation with Daryl. She watches the slow, dark river. She smiles at some secret, private thought. She takes out her sketchbook, in which she draws flowers. Her presence is a comfort, just the simple fact of her.

From Tortuguero, we go back on a boat to the biology field station in the Barra del Colorado Wildlife Refuge, where we are staying in rustic conditions. The current manager is a young man who warns us always to bring a flashlight when we use the outside bathroom, a favorite spot for nocturnal fer-de-lances. The fer-de-lance is a long, brown, venomous snake with a reputation for running at you instead of away. Ditto the long, brown, venomous bushmaster.

When we go for a walk, the manager tells us not to brush against certain trees which may shelter coral snakes, their mouths so small they mostly grab and bite the loose skin between human fingers. I comment that the pupae of a butterfly in this rainforest have evolved to mimic the head of a pit viper, with false scales, pits, and slitted eyes. The manager is not surprised and mentions that, in his year here, he has only had to rush one visitor to the hospital.

The walk we are taking is through knee-deep water that occasionally rises up to my thigh. The trail is flooded from the nearby rain-swollen canal, an event so common that the path is marked by red flags tied eye-level on the palm trees lacing the sky. The visual space in a rainforest is almost always filled, in this one by fronds, ferns, and the black rippling water. Periodically I slip on a submerged branch, as does my daughter.

The afternoon is very hot and very muggy. A cloud of mosquitoes evaluates the repellant we had applied earlier, hurriedly. On a drier stretch of land, we stop to watch a trail of army ants, the big soldiers with mandibles agape. The manager once saw these ants march through his field station, a rippling road three feet wide, through the kitchen and office and open dining area, eating all the dead beetles I have noticed on the floor, leaving the station spotlessly clean. Meanwhile, the manager moved out to a lodge across the canal.

Army ants will devour almost anything sluggish in their path, yet they detour around the larvae of the Owl butterfly, which emit a chemical defense from their neck glands. The last instars of an Owl are gigantic, long and fat and brutish, weighing more than half an ounce, their head capsules erupting in eight clubbed horns. The largest two horns curve back dramatically, like those of bighorn sheep.

These larvae feed together in groups, without cannibalizing, gentle giants next to tiny first instars.

At three and one-half inches, the adult Owl is the largest butterfly in Costa Rica. Its two unblinking eyespots on the under hindwing may be mimicking a predator, or they may be false targets for a predator's attack.

In this dark rainforest, the cleared area around the biology field station is full of light and space and flowers and, in effect, has become a butterfly house, full of butterflies. From the open window of our cabin, my daughter and I can watch the flight of blue-gray Owls, yellow-and-black swallowtails, and the intelligent *Heliconius,* their black wings dashed with orange and red.

If we were to stay long enough, we could time one of these Postmans as he traplines his flowers, regular as the eleven o'clock delivery. We could search for a nocturnal Postman roost, careful to avoid fer-de-lances. We could catch a female Postman to see how she extends her "stink clubs," small aromatic glands near the genitals. Since the glands of virgins have little smell, Larry Gilbert believes the odor is an anti-aphrodisiac transferred by the male to the female during mating, one more No Trespassing sign.

But we do not stay long enough. The biology station has many surprises, an orchid in bloom, a luminous moth, a jaguar's footprint. But not all the surprises are good ones, including the hundred mosquito bites that cover my daughter's legs and back.

In the 1930s, Evelyn Cheeseman was a butterfly collector who traveled to remote places such as Papua New Guinea, where local newspapers would exclaim with some disapproval: "Woman, 68, Is Jungle Prowling!" On one adventure, Lady Cheeseman confesses that she broke camp "owing to an untoward incident. For I found a leech in the teapot. The huts were fairly free of them, but specimens did appear from time to time carried in by ourselves inadvertently. This was a tortoise-shell leech which made it look still more repulsive: fat and sleek. It suddenly occurred to me that nobody but an idiot would remain in any place where there were leeches in the teapots; it would be super-human to go on any longer after this."

My daughter and I find our equivalent leech in the teapot, and we leave the biology field station in search of more and other butterflies, clearwings in the northern mountains, metalmarks on Pacific beaches, bright red *Mesenes* and speckled yellow *Agryrogrammanas,* checkered and swirled and long-tailed butterflies that have no common name but elicit only shouts of "Purple!" or "Lime green!"

We want all this, admittedly, in the context of a hotel.

In 1951, a dozen American Quaker families got in their cars and left the United States looking for a country without a military draft. Costa Rica had abolished its army some years before, and a later president would win the Nobel Peace Prize. As easy as that, the Friends settled in the Costa Rican mountain village of Monteverde and began dairy farming and cheese making. They put part of their land aside as a natural preserve, which expanded into the privately owned *Reserva Bosque Nubosa,* 26,000 acres of virgin tropical forest, often covered by low clouds. Other private reserves, including 32,000 acres of the Children's Eternal Forest, connect with a national park to create a large area of protected land.

Monteverde is a major site for ecotourists. Some landowners now find the nature business so profitable that they have let their coffee and banana plantations grow back into secondary rainforests, where they give daily and nightly tours. My daughter and I go on such a "twilight walk" in a guided group of eight, which keeps bumping into the four other guided groups of eight as, flashlights in hand, we visit the same roosting bats and leaf-cutter's nest, watching for semi-tame coatimundis, prehensile porcupines, tree frogs, sleeping birds, and sleeping butterflies.

Our guide feels the need to model reverence. "Notice how my voice changes," he whispers, "when I am in the forest." Later, after we fail to see the prehensile porcupine, he reminds us that life is unpredictable. Indeed, he says, "Life is a tour."

My twilight walk costs $14. Entrance into the Monteverde cloud forest is $12, a guided tour an additional $15 per person. Another private reserve features the Sky Walk, a series of suspension bridges above the forest canopy, and the Sky Trek, a jungle ride in which customers strap on a harness and swing fast through the air; you can pay separately or do both for $45. Entrance into the Santa Elena Reserve is $12 (proceeds go to the local high school) and a trail in the Children's Eternal Forest is $7. There is an excellent butterfly house, started and run by an American biologist, which also costs $7.

Of course, every park, every reserve, every butterfly house has a gift shop.

Certain butterflies are associated with cloud forest. Commonly I see clearwings, over an inch long, their wings like pieces of glass edged in brown. These butterflies fly low in the canopy, at eye level, winking in and out of existence. The males visit blue asters from which they absorb chemical compounds they use as a pheromone to attract females. The females skip about, visible, invisible, hiding in sunlight, laying their eggs.

Costa Ricans call the pupae of clearwings *espejitos* (little mirrors). My daughter lifts a leaf in the Monteverde cloud forest, and we see a half-dozen bright chrysalides bouncing with movement. We can't help ourselves. We both exclaim loudly.

Like ravens, we are drawn to glitter. We want to take these home to our nest.

In this rainforest, the overdecorated trees are festooned with other plants, orchids and ferns and countless vines. The trees drip rain and moss, glistening leaves, dazzle and shadow. For caterpillars and pupae, it is the same game here as everywhere else: Hide in plain sight, look like something else.

The larvae of a black-and-orange butterfly are pretending to be moss-covered twigs, while other species, more ambitious, impersonate moss itself or dead leaf matter. Some caterpillars are making frass chains, repositioning their waste to confuse predatory ants who avoid walking over thin strands of vegetation. Many larvae use countershading, their paler undersides helping them blend into the background.

Only a few stand out deliberately, black spines bristling. These spines deter predacious insects as well as any mammal or lizard who has had experience with highly toxic moth larvae. (White-faced monkeys are known to avoid even slightly hairy caterpillars; squirrel monkeys go through a complicated process of de-spining them.)

The path we are on leads to an overlook of pristine rainforest, the very tops of trees. Below me on the hillside are impatiens in a spectrum of pink, the triangular red-and-yellow bracts of heliconids, white calla lilies, passion vines, and the blue asters that attract male clearwings. Hummingbirds nectar with a flash of purple throat or green breast. Occasionally, the prong-billed barbet, or "squeaky gate bird," gives a metallic alarm. A tiny morpho escapes from the sky.

The business of nature comes down to this, this continuing wilderness, which is here only because people want it to be here. Costa Rica has the highest percentage of protected land in Central America; Costa Rica also has one of the highest rates of deforestation in Central America. It is all about business, what you sell, what you buy, and how much you are willing to pay.

Quakers are still very present in Monteverde. They own some of the hotels and restaurants and continue to operate a cheese-making factory. They run an alternative school for grades K–12, and they hold silent worship in their Meeting House every Sunday and Wednesday. I am a Quaker, and I have been looking forward to this Meeting, which begins at 10:00 A.M. with a circle of fifteen men, women, and children sitting on benches in a wooden house, singing traditional songs. At 10:30, silence begins. The children leave for a version of Sunday school. More adults come in. They sit down. No one says a word.

My Quaker Meeting in Silver City, New Mexico, also has silent worship, the tradition in which Friends sit and wait for something good to happen, for God, or for what we call the Light to make its Presence known. There is no sermon or ritual, and members rise to speak only when they feel powerfully called. Mostly, there is silence. Mostly, we sit and wait.

The mind tends to wander. I am thinking ahead to Business Meeting, which I won't be attending but which should come next. Because Quakers have no paid leader, we rely heavily on committees for our organizational work. Business Meetings can be tedious, although they are meant, as well, to be meetings for worship. The business of waiting on the Light and the business of the Meeting are meant to be the same, and we are all supposed to remember this as the committee reports get longer and longer and discussion continues over points so petty we want to stand on our wooden chairs and scream.

There is some analogy to the nature business.

Life is a tour.

In the United States, the business of butterflies includes commercial farms that raise Monarchs to be released during special occasions, such as weddings. For $65, plus $25 overnight shipping, I can buy a dozen Monarchs ready to fly. These farms also sell kits to schools and educators. A classroom will get a handful of Painted Lady larvae, along with a food supply. The children can watch the caterpillars eat, grow, molt, pupate, and turn into butterflies.

The controversy revolves around the butterfly's release. Most scientists are appalled by the idea of Painted Ladies from California breeding with Painted Ladies from Iowa. They worry about disease and the unnatural mingling of populations. They describe such releases as "ecological pollution." They are aesthetically displeased, as well.

Commercial breeders respond that they take precautions against disease. They dismiss the idea that a few released butterflies could affect wild populations. They say it's a wonderful business that helps teachers and children and brings people joy.

Currently in the United States, federal laws allow for the transport across state lines of nine species in their natural range, most popularly Monarchs, Painted Ladies, American Painted Ladies, and Red Admirals.

In 2001, one commercial farm shipped 82,000 Monarch butterflies and chrysalides. The cost was $3.50 per chrysalis and $95 for a dozen butterflies. Thirty-six thousand of these Monarchs were released at weddings.

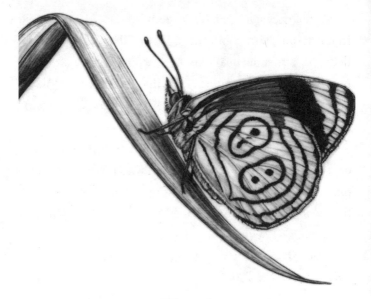

88 butterfly

Up to our very last moment in Costa Rica, on our last walk through the carpeted hallway that leads to our plane, I am looking for the 88 butterfly, named for the distinct number 88 written in brownish orange on the white background of the underside. The 88 butterfly, which can range as far north as southern Texas, is considered anthropophilic, flying into human houses and drawn inquisitively to human hair and clothes. Related species have other numbers on them, 69 or 68 or 89, with decorative patterns of stripes and circles in patterns of yellow, red, blue, blue-green, and orange.

Philip DeVries notes that this genus is commonly framed as a picture or covered with plastic and used to decorate place mats, coasters, and dinner plates. Those of us who squirm might remember that humans have always enjoyed natural objects as ornament, from shell buttons to feathers in our hair.

In Costa Rica, to have an 88 or 89 fly into your house is considered an extraordinary sign of good luck. The appropriate response is to run out immediately and buy a lottery ticket.

FIFTEEN

AIR AND ANGELS

WHY DO WE LOVE BUTTERFLIES?

Our pleasure often seems rooted in childhood. Beauty passes by on the wing. We are lucky if no one suggests, casually, that this is inconsequential, that beauty is just a passing thing, and the beauty of butterflies too brief, too frail to be of much use, that beauty is not power.

In any event, if we were to notice all the beauty of the world, what would become of us? The sunflower would stop us in our tracks, the cloud-tossed sky delay us for hours. We'd never get to school. We'd never get in the car.

We are lucky, in childhood, if we have good hearing: if we hear beauty and love and uselessness shouting in our ear, all the time, from every corner of the natural world.

We are lucky if, for just a moment, we feel that "radical interchange of separate identities." We are lucky if we can walk through secret dimensions as naturally as we walk from room to room.

Few of us would say that childhood is a simple time or a safe one. Certainly, butterflies are not about safety or lost happiness.

We grow to admire the caterpillar, that bag of goo, its blood like a clock ticking time. We watch it navigate a perilous world. We understand its fierceness, its need for deception. No matter our religious beliefs, we accept the miracle of metamorphosis. One thing becomes another. The Queen emerges, brilliant, heartbreaking, and we follow the course of her brief, obsessive days until her colors are faded and her wings torn. Now we lift up leaves, in search of eggs.

We have learned that beauty is not about comfort.

Because we are human, we probe the mystery. Genitalia have eyes. Swallowtails can remember. Evolution dodges to avoid a bat. Butterflies evolve ears on their wings. Wings pretend to be a head.

Evolution expresses itself so generously, in so many forms, and we become obsessive ourselves, wanting to know them all, to own them all, to put them in order. Like the gods in our myths, we name the creatures of the world: the White Admiral, the Mourning Cloak, the Silver-spotted Fritillary, the Great Copper, the Cloudless Sulphur, the Eastern Comma, the Field Crescent, the Paradise Birdwing, the Palos Verdes Blue, the Wood White, the Southern Festoon, the Purple Emperor, the Snout, the Brimstone.

We put them in drawers. We put them on our walls.

We could spend our lives counting butterflies.

Why do we love butterflies?

I think we have a physical response to color.

Flowers evolved color to attract the bee, the hummingbird, and the butterfly. Flowers, passionately, want to receive and send out pollen. *Come to me,* the flower shouts. Yellow is language. Purple is advertising.

Animals use the same strategy. *Look,* the gibbon says, *I have a big blue bottom.* The peacock spreads his ridiculous tail.

Color also warns. The red berry tastes bad. The gold beetle is poisonous.

Love and fear. Attract and repel. This all means something in the multitudinous greens and browns of the rainforest, the monotonous conifers and prairie grasses, the pastel desert. Color is an exclamation we once understood and still exploit. Look up from this page and you will see color, not greens and browns, but the red Coke can, the orange wallpaper, the pink dress, the purple hairbrush. Come to me! Buy me!

Inured as we are, surrounded by melodrama, we respond still to the flap of the Painted Lady, to the blue morpho.

The writer Annie Dillard says, "We teach our children one thing, as we were taught, to wake up."

Butterflies wake us up.

We are storytelling animals. We add a story. We add a thousand and one.

A single butterfly is a vain woman, a *geisha,* a fickle lover.

Two butterflies mean marital happiness.

Four butterflies are bad luck.

Red butterflies are witches.

The butterfly is the Creator who flew over the world searching for a place where humans could live.

At night, butterflies bring us dreams.

Rice has a butterfly soul.

Butterflies came from the tears of the Virgin Mary.

A butterfly will show you your true love.

Migrating sulphurs are pilgrims on their way to Mecca.

Butterflies are *hu dieh,* stemless flowers.

Butterflies are the souls of children.

Butterflies steal butter.

A dye of butterfly wings will make your pubic hair grow strong.

The spirit of the butterfly is in a Hopi kachina.

A black butterfly means death.

Hordes of butterflies predict famine.

White butterflies mean a rainy summer.
Butterflies bring the spring.
A man in love has butterflies in his belly.
Butterflies are stray, familiar thoughts.
Butterflies are air and angels.

Butterfly kachina

SELECTED BIBLIOGRAPHY
AND NOTES

A NOTE ON NAMES

The classification system is one described in a number of books, including Phil Schappert, *A World for Butterflies: Their Lives, Behavior, and Future* (Buffalo, N.Y.: Firefly Books, 2000); and in Rod and Ken Preston-Mafham, *Butterflies of the World* (London: Blandford Books, 1999), with the exception of the Riodinidae, which is listed as a family rather than a subfamily. A good case for that listing can be found in Philip DeVries, *The Butterflies of Costa Rica and Their Natural History,* vol. 2, *Riodinidae* (Princeton: Princeton University Press, 1997). The epigraph is from Alfred Russel Wallace, *The Malay Archipelago* (London: Macmillan and Company, 1869).

OBSESSION WITH BUTTERFLIES

An important book for any butterfly enthusiast in North America is James A. Scott, *The Butterflies of North America: A Natural History and Field Guide* (Stanford: Stanford University Press, 1986). I also found information on the Western Tiger Swallowtail in the Peterson Field Guide Series: Paul A. Opler, *A Field Guide to Western Butterflies,* illustrated by Amy Bartlett Wright (New York:

Houghton Mifflin, 1999), and in J. Mark Scriber, "Tiger Tales: Natural History of Native North American Swallowtails," *American Entomologist* (spring 1996).

Eleanor Glanville's quote on the fritillary's pupa comes from Ronald Sterne Wilkinson, "Elizabeth Glanville: An Early English Entomologist," *Entomologist's Gazette,* vol. 17 (October 1966), as does the quote from the "well-known entomologist."

A wonderful and well-researched book on British collectors is Michael Salmon, *The Aurelian Legacy: British Butterflies and Their Collectors* (Great Horkesley, Essex: Harley Books, 2000). From there I took the quote on the Glanville Fritillary, which was originally written by the Reverend J. F. Dawson in 1846. *The Aurelian Legacy* also tells the story of Eleanor Glanville and her disputed will, as does W. S. Bristowe, "The Life of a Distinguished Woman Naturalist, Eleanor Glanville (circa 1654–1709)," *Entomologist's Gazette,* vol. 18 (November 1966). Other sources include C. E. Goodricke, *The History of the Goodricke Family* (London, 1885); and P.B.M. Allan, "Mrs. Glanville and Her Fritillary," *Entomologist's Records Journal,* vol. 63 (1951).

A good book on the associations of religious and mythical figures with butterflies is Maraleen Manos-Jones, *The Spirit of Butterflies: Myth, Magic, and Art* (New York: Harry N. Abrams, 2000). I also used material from Miriam Rothschild, *Butterfly Cooing Like a Dove* (New York: Doubleday, 1991).

The quote on those "deprived of their Senses" is from Moses Harris, *The Aurelian or Natural History of English Insects, Namely Moths and Butterflies* (1766; reprint, Salem House Publishers, 1986).

The quote by David Allan is from his *The Naturalist in Britain: A Social History* (Princeton: Princeton University Press, 1976).

Information on the names and history of field clubs comes from Salmon, *The Aurelian Legacy*. Material on Lord Rothschild comes primarily from Miriam Rothschild, *Dear Lord Rothschild: Birds, Butterflies and History* (Glenside, Pa.: Balaban Publishers, 1983).

Information on and quotes by A. S. Meek come from his *A Naturalist in Cannibal Land* (London: Adelphi Terrace, 1913). The quote by Theodore Mead is from Grace Brown, ed., *Chasing Butterflies in the Colorado Rockies with Theodore Mead in 1871, as Told Through His Letters,* annotated by F. Martin Brown (Colorado Outdoor Education Center, Bulletin Number 3, 1996). The author of the guide on eastern butterflies in 1898 is Samuel Scudder.

The numbers concerning lepidoptera come from Phil Schappert's beautifully illustrated *A World for Butterflies: Their Lives, Behavior, and Future* (Buffalo, N.Y.: Firefly Books, 2000); other sources, such as Rod and Ken Preston-Mafham, *Butterflies of the World* (London: Blandford Books, 1999), have slightly different numbers (160,000 species of lepidoptera with 20,000 being butterflies).

The quote from Chuang Tze is well-known. I took my version from Manos-Jones, *The Spirit of Butterflies*. The modern interpreter of Chuang Tze is Kuang-ming Wu and the quotes come from his *The Butterfly as Companion: Meditations on the First Three Chapters of the Chuang Tze* (New York: State University of New York Press, 1990). The quote from Marcel Roland comes from a translation by Judith Landry of his *Vues sur le monde animal: Amour, harmonie, beauté,* published in 1943.

Material on Miriam Rothschild comes from her *Dear Lord Rothschild*, as well as Salmon, *The Aurelian Legacy*. The brief discussion of her work and her quote come from her essays in *Butterfly Gardening: Creating Summer Magic in Your Garden* (San

Francisco: Sierra Club Books, 1998), which is also the source for the last quote in this chapter.

The material and quotes from John Tennent are from personal correspondence.

TOUGH LOVE

Bert Orr is the source for some of the images in this chapter, such as the squashed golf ball and the skipper larva rearing up like a cobra.

General information on the biology of larvae can be found in Malcolm Scoble, *The Lepidoptera: Form, Function, and Diversity* (New York: Oxford University Press, 1992); Amy Bartlett Wright, *Peterson's First Guide to Caterpillars* (Boston: Houghton Mifflin, 1993); James Scott, *The Butterflies of North America: A Natural History and Field Guide* (Stanford: Stanford University Press, 1986); and Philip DeVries, *The Butterflies of Costa Rica and Their Natural History,* vol. 2, *Riodinidae* (Princeton: Princeton University Press, 1997). The estimate that some caterpillars gain 3,000 times their hatching weight comes from Phil Schappert, *A World for Butterflies: Their Lives, Behavior, and Future* (Buffalo, N.Y.: Firefly Books, 2000).

More information about swallowtails can be found in J. Mark Scriber, Yoshitaka Tsubaki, and Robert Lederhouse, eds., *Swallowtail Butterflies: Their Ecology and Evolutionary Biology* (Gainesville, Fla.: Scientific Publishers, 1995).

More information about the defenses of caterpillars and of plants can be found in Nancy Stamp and Timothy Casey, eds., *Caterpillars: Ecological and Evolutionary Constraints on Foraging* (London: Chapman and Hall, 1993). Particularly useful essays in this book are David Dussord, "Foraging with Finesse: Caterpillar Adaptations for Circumventing Plant Defenses"; Bernd Heinrich, "How Avian Predators Constrain Caterpillar Foraging"; M. Deane Bowers, "Aposematic Caterpillars: Life-Styles of the Warningly Colored and Unpalatable"; and Nancy Stamp and Richard Wilkens, "On the Cryptic Side of Life: Being Unapparent to Enemies and the Consequences for Foraging and Growth of Caterpillars."

The material on caterpillar locomotion comes primarily from John Brackenbury, "Fast Locomotion in Caterpillars," *Journal of Insect Physiology,* vol. 45 (1999).

The experiment with wasps and Asian swallowtails is described by Masami Takagi et al., "Antipredator Defense in *Papilio* Larvae: Effective or Not?" in Scriber, Tsubaki, and Lederhouse, *Swallowtail Butterflies.*

The architecture of skipper larvae is discussed in Martha Weiss et al., "Ontogenetic Changes in Leaf Shelter Construction by Larvae of *Epargyreus Clarus* (Hesperidae), the Silver-spotted Skipper," *Journal of the Lepidoptera Society,* vol. 54, no. 3 (2001). Information about the ejection of frass also came from personal correspondence with Martha as well as from May Berenbaum, "Shelter-Building Caterpillars: Rolling Their Own," *Wings: Essays on Invertebrate Conservation* (Portland, Ore.: The Xerces Society, fall 1999).

The quote from Miriam Rothschild comes from the essay by David Dussord in Stamp and Casey, *Caterpillars.*

More information on chemical defenses in plants and signaling among plants and insects can be found in my *Anatomy of a Rose* (Cambridge, Mass.: Perseus, 2001), and in my discussion of Ian Baldwin's work in "Talking Plants," *Discover Magazine,* vol. 23, no. 4 (April 2002); this article includes work by Consuelo DeMoraes et al., described in the article "Caterpillar-Induced Nocturnal Plant Volatiles Repel Conspecific Females," *Nature,* vol. 410 (March 2001). The main source for the theory concerning bacteria in a caterpillar's gut is the article by Wilhelm Bolland et al., "Gut Bacteria May Be Involved in Interactions Between Plants, Herbivores, and Their Predators," *Biological Chemistry*, vol. 381 (August 2000).

YOU NEED A FRIEND

The material on Philip DeVries, ants, and butterflies is from personal correspondence, as well as the following works by Phillip DeVries: *The Butterflies of Costa Rica and Their Natural History*, vols. 1 and 2 (Princeton: Princeton University Press, 1997), and "Singing Caterpillars, Ants and Symbiosis," *Scientific American* (October 1992). The species name of the metalmark described is *Thisbe irenea*.

Material on the Australian Bright Copper comes from J. Hall Cushman et al., "Assessing Benefits to Both Participants in a Lycaenid-Ant Association," *Ecology*, vol. 75, no. 4 (1994). Material on the Common Imperial Blue is from Mark Travassos and Naomi Pierce, "Acoustics, Context, and Function of Vibrational Signaling in a Lycaenid Butterfly Ant Mutualism," *Animal Behavior*, vol. 60 (2000). Information on the carnivorous blue larva that eats aphids is from personal correspondence with Bert Orr. Material on the European blue, which resembles a monstrous ant grub, is from J. C. Wardlaw et al., "Do *Maculinea rebeli* Caterpillars Provide Vestigial Mutualistic Benefits to Ants When Living as Social Parasites Inside *Myrmica* Ant Nests?" *Entomologia Experimentalis et Applicata*, vol. 95 (2000). I referred to other articles as well, including Thomas Damm et al., "Adoption of Parasitic *Maculinea Alcon* Caterpillars by Three *Myrmica* Ant Species," *Animal Behavior*, vol. 62 (2001).

The problems and natural history of the English Large Blue are discussed in Phil Schappert, *A World for Butterflies: Their Lives, Behavior, and Future* (Buffalo, N.Y.: Firefly Books, 2000), and John Feltwell, *The Natural History of Butterflies* (London: Facts on File Publications, 1986). The quote from Vladimir Nabokov comes from *The Gift,* translated by Michael Scammel in 1952, quoted in Robert Michael Pyle, *Nabokov's Butterflies: Unpublished and Uncollected Writings* (Boston: Beacon Press, 2000). The quotes from Sir Compton McKenzie are in Patrick Matthews, *The Pursuit of Moths and Butterflies: An Anthology* (London: Chatto and Windus, 1957).

Information on the Madrone caterpillar comes primarily from Terrence D. Fitzgerald, "Nightlife of Social Caterpillars," *Natural History* (February 2001). General information on the longevity of caterpillars, social caterpillars, and signals for molting can be found in James Scott, *The Butterflies of North America: A Natural History and Field Guide* (Stanford: Stanford University Press, 1986), and Malcolm Scoble, *The Lepidoptera: Form, Function, and Diversity* (New York: Oxford University Press, 1992).

In *The Natural History of Butterflies,* John Feltwell mentions the role of carotenoid pigments in the yellow blood of caterpillars and the ability of the Large White to count hours of light.

METAMORPHOSIS

An excellent book on Vladimir Nabokov is Robert Michael Pyle, *Nabokov's Butterflies: Unpublished and Uncollected Writings* (Boston: Beacon Press, 2000). The lecture quoted can be found in this book and was originally given in March 1951 at Cornell University in a Masterpieces of European Fiction class. Other sources for Nabokov's work and life are Vladimir Nabokov, *Speak, Memory:*

An Autobiography Revisited (New York: Putnam, 1966), and Kurt Johnson and Steven Coates, *Nabokov's Blues: The Scientific Odyssey of a Literary Genius* (Cambridge, Mass.: Zoland Books, 1999).

More information on metamorphosis can be found in general books by James Scott and Malcolm Scoble, and in H. Frederik Nijhout, *The Development and Evolution of Butterfly Wing Patterns* (Washington, D.C.: Smithsonian Institution Press, 1991).

The folktale concerning the Hindu god Brahma is a common one. The story of Pope Gelasius I, as well as other mythic facts and tidbits, are collected in Maraleen Manos-Jones, *The Spirit of Butterflies: Myth, Magic, and Art* (New York: Harry N. Abrams, 2000), and in Miriam Rothschild, *Butterfly Cooing Like a Dove* (New York: Doubleday, 1991). The material on convicts in China is from the Web site True Buddha School Net at www.tbsn.org/ebooks/satira/convicts.htm.

The quote from Elizabeth Kubler-Ross is from her memoir *The Wheel of Life: A Memoir of Living and Dying* (New York: Touchstone Books, 1997). The quote from Miriam Rothschild is from her *Butterfly Cooing Like a Dove*. Material on the Aztec relationship with butterflies can be found in many sources, including Laurette Sejourne, *Burning Water: Thought and Religion in Ancient Mexico* (Berkeley: Shambhala Books, 1976). Philip DeVries's quote comes from his *The Butterflies of Costa Rica,* vol. 1 (Princeton: Princeton University Press, 1997). Information on the emergence of the adult butterfly can be found in the general books already listed.

BUTTERFLY BRAINS

Material on Martha Weiss and her work comes from personal correspondence, as well as the following articles by Martha Weiss: "Innate Colour Preference and Flexible Colour Learning in the

Pipevine Swallowtail," *Animal Behavior*, vol. 53 (1997); "Brainy Butterflies," *Natural History*, vol. 109, no. 6 (July/August 2000); "Ontogenetic Changes in Leaf Shelter Construction by Larvae of *Epargyreus Clarus* (Hesperidae), the Silver-spotted Skipper," *Journal of the Lepidoptera Society*, vol. 54, no. 3 (2001); and Martha Weiss and Dan Papaj, "Colour Learning in Two Behavioral Contexts: How Much Can a Butterfly Keep in Mind?" (manuscript in preparation). Other sources include Susan Milius, "How Bright Is a Butterfly?" *Science News*, vol. 153 (11 April 1998); Dave Goulson et al., "Foraging Strategies in the Small Skipper Butterfly, *Thymelicus favus*: When to Switch?" *Animal Behavior*, vol. 53 (1997); C. M. Penz and H. W. Krenn, "Behavioral Adaptations to Pollen-Feeding in *Heliconius* Butterflies," *Journal of Insect Behavior*, vol. 13, no. 6 (2000); and Camille McNeely and Michael Singer, "Contrasting the Roles of Learning in Butterflies Foraging for Nectar and Oviposition Sites," *Animal Behavior*, vol. 61 (2002).

Quotes and information from Dan Papaj come from personal correspondence, as well as some of the articles already noted.

Information on bees comes from various sources, including Frederich Barth's *Insects and Flowers* (Princeton: Princeton University Press, 1991).

BUTTERFLY MATISSE

More information on the design and colors of butterfly wings can be found in the general books already noted, as well as Rod and Ken Preston-Mafham, *Butterflies of the World* (London: Blandford Books, 1999), and H. Frederik Nijhout, *The Development and Evolution of Butterfly Wing Patterns* (Washington, D.C.: Smithsonian Institution Press, 1991). That book is the source of the quote by Nijhout.

The experiment with sulphurs is described in Richard Vane-Wright and Michael Boppre, "Visual and Chemical Signaling in Butterflies: Function and Phylogenetic Perspectives," *Phil. Trans. Royal Society of London,* vol. 340 (1993).

Information on the African butterfly, whose species name is *Bicyclus anynana,* comes from Sean Carroll, "Genetics on the Wing: Or How the Butterfly Got Its Spots," *Natural History,* vol. 2 (1997), and Paul Brakefield et al., "The Genetics and Development of an Eyespot Pattern in the Butterfly *Bicyclus anynana*: Response to Selection for Eyespot Shape," *Genetics,* vol. 46 (May 1997), as well as other articles on *Bicyclus* butterflies by Paul Brakefield. This African species is not to be confused with the African species *Precis octavia,* whose wet-season form is blue and dry-season form orange-red.

LOVE STORIES

Love stories among butterflies can be found in the general books already noted. N. Tinbergen first described the Grayling's courtly bow in *The Study of Instinct* (Folcroft, Pa.: Folcroft Press, 1951). I also read Robert Lederhouse, "Comparative Mating Behavior and Sexual Selection in North American Swallowtail Butterflies" and Kazuma Matsumoto and Nobuhiko Suzuki, "The Nature of Mating Plugs and the Probability of Reinsemination in Japanese Papilionidae," both in J. Mark Scriber, Yoshitaka Tsubaki, and Robert Lederhouse, eds., *Swallowtail Butterflies: Their Ecology and Evolutionary Behavior* (Gainesville, Fla.: Scientific Publishers, 1995), as well as Darrell J. Kemp and Christer Wiklund, "Fighting Without Weaponry: A Review of Male-Male Contest Competition in Butterflies," *Behavorial Ecology Sociobiology,* vol. 49 (2001).

Material on the eyes in a swallowtail's genitalia can be found in the following articles by Kentaro Arikawa: "Hindsight of Butterflies," *Bioscience,* vol. 51, no. 3 (March 2001), and "The Eyes Have It," *Discover Magazine,* vol. 17 (November 1996).

More information on chemical signaling in milkweed butterflies and their use of alkaloids can be found in Richard Vane-Wright and Michael Boppre, "Visual and Chemical Signaling in Butterflies: Function and Phylogenetic Perspectives," *Phil. Trans. Royal Society of London,* vol. 340 (1993), and in Michael Boppre, "Sex, Drugs, and Butterflies," *Natural History,* vol. 103 (January 1994). An important book on these butterflies is P. R. Ackery and R. I Vane-Wright, *Milkweed Butterflies: Their Cladistics and Biology* (Ithaca: Cornell University Press, 1984). The observation of monarchs drinking dew is from Susan Milius, "Male Butterflies Are Driven to Drink," *Science News* (24 August 2002).

Mud-puddling is discussed in many sources, including Carol Boggs and Lee Ann Jackson, "Mud-Puddling by Butterflies Is Not a Simple Matter," *Ecological Entomology,* vol. 16 (1991).

Pupal mating is also mentioned in many books and articles, including Larry Gilbert, "Biodiversity of a Central American *Heliconius* Community: Pattern, Process, and Problems," in *Plant-Animal Interactions: Evolutionary Ecology in Tropical and Temperate Regions* (New York: Wiley and Sons, 1991).

Material on the sphragis comes from personal correspondence with Bert Orr, as well as his chapter "The Evolution of the Sphragis in the Papilionidae and Other Butterflies," in Scriber, Tsubaki, and Lederhouse, *Swallowtail Butterflies.* I also used A. G. Orr and Ronald Rutowski, "The Function of the Sphragis in *Cressida Cressida,*" *Journal of Natural History,* vol. 25 (1991), and A. G. Orr, "The Sphragis of *Heteronympha penelope* Waterhouse: Its Structure, Formation and Role in Sperm Guarding," *Journal of Natural History,* vol. 36 (2002). The butterfly that produces an internal stalk is *Acraea natalica.*

More information on the Cabbage White can be found in Johan Anderson et al., "Sexual Cooperation and Conflict in Butterflies: A Male-Transferred Anti-Aphrodisiac Reduces Harassment of Recently Mated Females," *Proceedings of the Royal Society of London,* vol. 267 (2001).

THE SINGLE MOM

More material on oviposition can be found in J. Mark Scriber, Yoshitaka Tsubaki, and Robert C. Lederhouse, eds., *Swallowtail Butterflies: Their Ecology and Evolutionary Behavior* (Gainesville, Fla.: Scientific Publishers, 1995), particularly in the following chapters: Mark Rausher, "Behavorial Ecology of Oviposition in the Pipevine Swallowtail, *Battus Philenor*"; Yoshitaka Tsubaki, "Clutch Size Adjustment by *Luehdorfia Japonica*"; and Ritsuo Nishida, "Oviposition Stimulants of Swallowtail Butterflies." Dan Papaj also provided information through personal correspondence. In addition, I used a variety of articles, including Camille McNeeley and Michael Singer, "Contrasting the Roles of Learning in Butterflies Foraging for Nectar and Oviposition Sites," *Animal Behavior,* vol. 61 (2001). The species name for Texas Dutchman's pipe is *Aristolochia reticulata;* the species name for Virginia snakeroot is *Aristolochia serpentaria.*

ON THE MOVE

Material on the migration of Snouts is taken from personal correspondence with Larry Gilbert, as well as his "Ecological Factors Which Influence Migratory Behavior in Two Butterflies of the Semi-Arid Shrublands of South Texas," *Contributions in Marine Science,* vol. 27 (Austin, Tex.: Marine Science Institute, University of Texas at Austin, September 1985). I also consulted Charles Gable and W. A. Baker, "Notes on a Migration of *Libythea bachmanni," The Canadian Entomologist,* vol. 12 (December 1922).

The quote from Vladimir Nabokov is from his memoir *Speak, Memory: An Autobiography Revisited* (New York: Putnam, 1966).

Material on the Painted Lady can be found in general sources, as well as Derham Giuliani and Oakley Shields, "Large scale Migra tions of the Painted Lady Butterfly, *Vanessa cardui,* in Inyo County, California, During 1991," *Bulletin of Southern California Academic Sciences,* vol. 94, no. 2 (1995). The quote on the migration of Painted Ladies on the Sudanese Red Coast was taken from Torben Larsen, "Butterfly Mass Transit," *Natural History,* vol. 102 (June 1993). In this article, Larsen also wrote about migrating butterflies willing to "batter down the house" as they flew straight to their goal.

Material on Monarchs comes from many general sources. I recommend Sue Halpern, *Four Wings and a Prayer* (New York: Pantheon Books, 2001), and Lincoln Brower, "New Perspectives on the Migration Ecology of the Monarch Butterfly," *Contributions in Marine Science,* vol. 27 (Austin, Tex.: Marine Science Institute, University of Texas at Austin, September 1985). Material on navigation in Monarchs comes from the following articles by Sandra Perez et al.: "A Sun Compass in Monarch Butterflies," *Nature,* vol. 387 (May 1997), and "Monarch Butterflies Use a Magnetic Compass for Navigation," *Proceedings of the National Academy of Sciences at the United States of America,* vol. 96, no. 24 (23 November 1999); and from Laura Tangley, "Butterfly Compasses," *US News and World Report,* vol. 127, no. 22 (6 December 1999), as well as from other articles.

For more information on migration, I also read Robert Srygley, "Compensation for Fluctuations in Crosswind Drift Without Stationary Landmarks in Butterflies Migrating Over Seas," *Animal Behavior,* vol. 61 (2001); Ilkka Hanski et al., "Metapopulation Structure and Migration in the Butterfly *Melitaea cinxia,*" *Ecology,* vol. 75, no. 3 (1994); Constanti Stefanescu, "The Nature of Migration in the Red Admiral Butterfly, *Vanessa atalanta*: Evidence from the Population Ecology in Its Southern Range," *Ecological Entomology,* vol. 26 (2001); and the following articles by Thomas Walker: "Butterfly Migrations in Florida: Seasonal Patterns and Long-Term Change," *Environmental Entomology,* vol. 30, no. 6 (December 2001), and "Butterfly Migration from and to Peninsular Florida," *Ecological Entomology,* vol. 16 (1991).

IN THE LAND OF BUTTERFLIES

More on the life of Henry Walter Bates can be found in George Woodcock, *Henry Walter Bates: Naturalist of the Amazons* (London: Faber and Faber, 1969). Most of the quotes by Bates are from his *The Naturalist on the River Amazon* (London: John Murray, 1876). The description of his attire, however, is from his "Proceedings of Natural History Collectors in Foreign Countries," *The Zoologist,* vol. 15 (1857). A good article on his collecting adventures and techniques in the field is Kim Goodyear and Philip Ackery, "Bates, and the Beauty of Butterflies," *The Linnean,* vol. 18 (2002). The quote concerning entomologists being "a poor set" comes from that article. I also quote from Bates's lecture "Contributions to an Insect Fauna of the Amazon Valley," read before the Linnean Society on 21 November 1861.

Information on mimicry can be found in many general sources. Phil Schappert, *A World for Butterflies: Their Lives, Behavior, and Future* (Buffalo, N.Y.: Firefly Books, 2000), has a good description and illustration of mimicry rings. I also consulted James Marden, "Newton's Second Law of Butterflies," *Natural History,* vol. 1 (1992); H. Frederick Nijhout, "Developmental Perspectives in Evolution of Butterfly Mimicry," *Bioscience,* vol. 44, no. 3 (March 1994); Peng Chai and Robert Srygley, "Predation and the Flight, Morphology, and Temperature of Neotropical Rainforest Butterflies," *The American Naturalist,* vol. 135, no. 6 (June 1990); Larry Gilbert and James Mallet, "Why Are There So Many Mimicry Rings? Correlations Between Habitat, Behavior, and Mimicry in *Heliconius* Butterflies," *Biological Journal of the Linnean Society,* vol. 55 (1995); Peng Chai and James Marden, "Aerial Predation and Butterfly Design: How Palatability, Mimicry, and the Need for Evasive Flight Constrain Mass Allocation," *The American Naturalist,* vol. 158, no. 1 (July 1991); David Ritland, "Variation in Palatability of Queen Butterflies and Implications Regarding Mimicry," *Ecology,* vol. 75, no. 3 (1994); Angus MacDougall and Marian Stamp Sawkins, "Predator Discrimination Error and the Benefits of Mullerian Mimicry," *Animal Behavior,* vol. 55 (1998); Robert Srygley and C. P. Ellington, "Discrimination of Flying Mimetic, Passion-Vine Butterflies *Heliconius,*" *Proceedings of the Royal Society of London,* vol. 266 (1999), David Ritland and Lincoln Brower, "The Viceroy Butterfly Is Not a Batesian Mimic," *Nature,* vol. 350 (11 April 1991); Richard Vane-Wright, "A Case of Self-Deception," *Nature,* vol. 350 (11 April 1991); and David Kapan, "Three-Butterfly System Provides a Field Test of Mullerian Mimicry," *Nature,* vol. 409 (18 January 2001).

Larry Gilbert's theory comes from his "Biodiversity of a Central American *Heliconius* Community: Pattern, Process, and Problems," *Plant-Animal Interactions: Evolutionary Ecology in Tropical and Temperate Regions* (New York: Wiley and Sons, 1991).

Bert Orr provided me with information in personal correspondence.

The final quote from Alfred Russel Wallace is from his *My Life: A Record of Events and Opinions* (New York: Dodd, Mead, and Company, 1906).

THE NATURAL HISTORY MUSEUM

Most of the material in this chapter comes from personal interviews with Jeremy Holloway, Richard Vane-Wright, Phil Ackery, and David Carter.

I also read J. D. Holloway and N. E. Stork, "The Dimensions of Biodiversity: The Use of Invertebrates as Indicators of Human Impact," *The Biodiversity of Microorganisms and Invertebrates: Its Role in Sustainable Agriculture,* ed. D. L. Hawksworth (CAB International, 1991); Richard Vane-Wright, "Taxonomy, Methods Of," in *Encyclopedia of Biodiversity,* vol. 3 (San Diego, Calif.: Academic Press, 2001); David Carter and Annette Walker, *Care and Conservation of Natural History Collections* (Newton, Mass.: Butterworth-Heinemann, 1997); "The Diversity of Moths: An Interview with J. D. Holloway," *Malayan Naturalist,* vol. 51, no. 1 (August 1997); Kim Goodyear and Philip Ackery, "Bates, and the Beauty of Butterflies," *The Linnean,* vol. 18 (2002); and Phil Ackery, "The Lepidoptera Collections at the Natural History Museum (BMNH) in South Kensington, London," *Holarctic Lepidoptera,* vol. 6, no. 1 (1999).

More material on the history of the museum can be found at the Web site of the Natural History Museum, as well as in John

Thackery and Bob Press, *The Natural History Museum: Nature's Treasurehouse* (London: Natural History Museum, 2001); and Mark Girouard, *Alfred Waterhouse and the Natural History Museum* (London: Natural History Museum, 1981).

The quotes about eating insects come from Vincent M. Holt, *Why Not Eat Insects?* (1885; reprint, London: Natural History Museum, 1967).

The brief quotes on age come from Edward O. Wilson, "A Grassroots Jungle in a Vacant Lot," *Wings: Essays on Invertebrate Conservation* (Portland, Ore.: The Xerces Society, fall 1995); Miriam Rothschild, "Ages Five to Fifteen: Wildflowers, Butterflies, and Frogs," by Miriam Rothschild in *Wings: Essays on Invertebrate Conservation* (Portland, Ore.: The Xerces Society, fall 1995); and Robert Michael Pyle, *The Thunder Tree: Lessons from an Urban Wildland* (Boston: Houghton Mifflin, 1993).

The quote from the naked collector is by Charles Morris Woodford and can be found in his *A Naturalist Among the Headhunters* (London: George Philip and Son, 1890).

The quote by Eleanor Glanville is from Ronald Sterne Wilkinson, "Elizabeth Glanville, an Early English Entomologist," *Entomologist's Gazette,* vol. 17 (October 1966).

NOT A BUTTERFLY

A good source of information about moths is Mark Young, *The Natural History of Moths* (London: T&AD Poyser Natural History, 1997), and Charles Covell, *A Field Guide to the Moths of Eastern North America* (Boston: Houghton Mifflin, 1984).

I also read Frederich G. Barth, *Insects and Flowers: The Biology of a Partnership* (Princeton: Princeton University Press, 1991); Michael Robinson, "An Ancient Arms Race Shows No

Sign of Letting Up," *Smithsonian,* vol. 23, no. 1 (April 1992); Richard Connif, "Purple, Orange, Oooh, He's Oozing Poison at Me," *Smithsonian,* vol. 26, no. 11 (February 1966); Darlyne Murawski, "Moths Come to Light," *National Geographic,* vol. 191, no. 3 (1997); Susan Milius, "Butterfly Ears Suggest a Bat Influence," *Science News,* vol. 157, no. 4 (22 January 2000); and Jens Rydell, "Echolocating Bats and Hearing Moths: Who Are the Winners?" *Oikos,* vol. 73, no. 3 (1995).

TIMELINE

Much of the material in this chapter comes from personal interviews and correspondence with Rudi Mattoni. More information can also be found in Rudi Mattoni, "The Endangered El Segundo Blue Butterfly," *Journal of Research on the Lepidoptera,* vol. 29, no. 4 (1990); Rudi Mattoni et al., "Analysis of Transect Counts to Monitor Population Size in Endangered Insects," *Journal of Insect Conservation,* vol. 5 (2002); Leslie Mieko Yap, "Brightening a Butterfly's Future," *National Wildlife* (October/November 1993); Rudi Mattoni and Travis Longcore, "Arthropod Monitoring for Fine-scale Habitat Analysis: A Case Study of the El Segundo Sand Dunes," *Environmental Management,* vol. 25, no. 4 (2000); Rudi Mattoni, "Rediscovery of the Endangered Palos Verdes Blue Butterfly, *Glaucopsyche lygdamus palosverdesensis,*" *Journal of Research on the Lepidoptera,* vol. 31, nos. 3–4 (1992); Connie Isball, "Green Teens Save the Blues," *Audubon* (September/October 1996); and Rudi Mattoni and Nelson Powers, "The Palos Verdes Blue: An Update," *Endangered Species Bulletin* (November/December 2000).

Information on Arthur Bonner comes from personal communications, as well as Michael Lipton, "Butterfly Man," *People Weekly,* vol. 49 (26 January 1998); *America's Endangered Species,* a National Geographic Special originally aired 24 January 1996 on NBC; and Tom Dworetzky, "In Helping Save Endangered Species, He Also Saved Himself," *National Wildlife,* vol. 35 (October/November 1997).

THE BUSINESS OF BUTTERFLIES

Bill Toone of the San Diego Museum gave me good background information, as did Daryl Loth of Tortuguero, Costa Rica. I also read Brent Davies, "Field Notes from a Costa Rican Butterfly Farm" and Bill Toone, "How a Bird Man Became a Butterfly Farmer in Costa Rica," both articles in *Wings: Essays on Invertebrate Conservation* (Portland, Ore.: The Xerces Society, spring 1995).

The quote from Miram Rothschild comes from her *Butterfly Cooing Like a Dove* (New York: Doubleday, 1991). The material from Philip DeVries comes from his *The Butterflies of Costa Rica and Their Natural History,* vol. 1 (Princeton: Princeton University Press, 1997).

More information on butterfly houses can be found in Robert Lederhouse et al., "Butterfly Gardening and Butterfly Houses and Their Influence on Conservation in North America," in J. Mark Scriber, Yoshitaka Tsubaki, and Robert C. Lederhouse, eds., *Swallowtail Butterflies: Their Ecology and Evolutionary Biology* (Gainesville, Fla.: Scientific Publishers, 1995).

More material on butterfly conservation in Papua New Guinea is in Larry Orsak, "Killing Butterflies . . . to Save Butterflies," on the Web site www.aa6g.org/Butterflies/pngletter.html; Michael Parsons, "Butterfly Farming and Trading in the Indo-Australian Region and Its Benefits in the Conservation of Swallowtails and Their Tropical Forest Habitats," in Scriber, Tsubaki, and Lederhouse, *Swallowtail Butterflies;* Thomas Hanscom, "Papua New Guinea: A

Butterfly Farming Success Story," *Wings: Essays on Invertebrate Conservation* (Portland, Ore.: The Xerces Society, spring 1995); and Michael Parsons, "Butterfly Conservation and Commerce in Papua New Guinea," in *The Butterflies of Papua New Guinea: Their Systematics and Biology* (San Diego, Calif.: Academic Press, 1999). Butterfly ranching in Kenya is discussed in Don Borough, "On the Wings of Hope," *International Wildlife,* vol. 30, no. 4 (July/August 2000), as well as in other articles.

Information on CRES is from personal correspondence with Joris Brinckerhoff and from his Web site, www.butterflyfarm.co.cr.

More material on the Barra del Colorado biology field station can be found at the COTREC Web site.

The quote by Evelyn Cheeseman is from her *Hunting Insects in the South Seas* (London: Philip Allan and Company, 1948).

A discussion on the commercial release of butterflies is in Judith Kirkwood, "Do Commercial Butterfly Releases Pose a Threat to Wild Populations?" *National Wildlife,* vol. 37, no. 1 (December 1998/January 1999), and June Kronholz, "Butterflies Are Free? Well, Not Under Rules Lepidopterists Debate," *Wall Street Journal,* 14 January 2002.

AIR AND ANGELS

The association of cultural and religious ideas with butterflies comes from many different sources, including Maraleen Manos-Jones, *The Spirit of Butterflies: Myth, Magic, and Art* (New York: Harry N. Abrams, 2000); Miriam Rothschild, *Butterfly Cooing Like a Dove* (New York: Doubleday, 1991); and numerous articles and Web sites. The idea of butterflies as "air and angels" and "stray, familiar thoughts" is repeated from quotes mentioned earlier in Chapter 1.

INDEX